The Future of
World Oil

The Future of World Oil

Paul Leo Eckbo

Ballinger Publishing Company • Cambridge, Massachusetts
A Subsidiary of J.B. Lippincott Company

338.27282
E19f

 This book is printed on recycled paper.

Library of Congress Cataloging in Publication Data

Eckbo, Paul Leo.
  The future of world oil.

  Originally presented as the author's thesis, Massachusetts Institute of Technology, 1975.
  1. Petroleum industry and trade. 2. Petroleum products—prices.
3. Trusts, Industrial. I. Title.
HD9560.4.E22 1976   338.2'7'282   76-16809
ISBN 0-88410-455-9

Copyright © 1976 by Ballinger Publishing Company. All rights reserved. No part of this publication may be reproduced, stored in a retrieval system, or transmitted in any form or by any means, electronic mechanical photocopy, recording or otherwise, without the prior written consent of the publisher.

International Standard Book Number: 0-88410-455-9

Library of Congress Catalog Card Number: 76-16809

Printed in the United States of America

Contents

v

APR 12 '79

HUNT LIBRARY
CARNEGIE-MELLON UNIVERSITY

List of Figures

List of Tables

Preface and Acknowledgments

This study was conducted and written during the years 1973 and 1975. The point of departure was the major petroleum discoveries made in Norway, of which the writer is a citizen, and the potential of the petroleum sector to become the most important sector of the Norwegian economy. The Norwegian economy would consequently become dependent on the international petroleum market. The international price of petroleum is, however, determined in a most complex fashion owing to the noncompetitive behavior in this market. This book attempts to develop a framework to analyze the implications of the likely degree of noncompetitive behavior in the international petroleum market. The focus is on the market strategies that may be pursued by the world's oil exporters on a joint or individual basis. The primary target of analysis is thus the price-making power of OPEC and the implications for the world's oil/energy markets should OPEC continue to exercise this power.

The original report on this study was submitted to the Alfred P. Sloan School of Management in September 1975 as a doctoral thesis in partial fulfillment of the requirements for the degree of Doctor of Philosophy at the Massachusetts Institute of Technology. The Ford Foundation, through its European Doctoral Fellowship Program in Management Education, provided the financial support for me to enter this doctoral program.

Early financial support was generously provided by Johan Peter Martens Handelshoyskolefond through the Center of Applied Research at the Norwegian School of Economics and Business Administration (SAF-NHH). Later work was financed by the "Analysis of the World Oil Market" project, which was initiated by the M.I.T. Energy Laboratory in association with the M.I.T. Sloan School of Management and the Department of Economics in the summer of 1974. The project was sustained at a modest level of effort by Energy Laboratory internal funds until March 1975, when support was received from the National Science Foundation (Grant No. SIA75-00739). The review of the previous commodity cartels draws on work done for the Department of the Interior under Contract No. P. O. A8309.

Professors M.A. Adelman and H.D. Jacoby have been the major source of inspiration and encouragement throughout this work. Professor Adelman introduced me to the intricacies of the petroleum industry and allowed me to find promising areas of exploration. Professor Jacoby has closely supervised the evolution of earlier drafts of this manuscript. His guidance and his emphasis on clarity and conciseness substantially improved both the content and the form of this manuscript.

Professor R.D. Robinson, the third member of the supervising committee, has been helpful and influential in my professional training. His teaching sparked my interest in studying the institutions, the governments, and the companies dominating international markets.

I am also thankful to Professor P.W. MacAvoy, whose guidance made it possible to explore meaningfully the experience of previous international commodity cartels.

The cost of this work project has been borne as much by my wife, Irmelin Munck, as by myself. Her encouragement during three long years was sorely needed. Her confidence was an inexhaustible source of inspiration, as was her description of what life could be beyond the completion of the thesis.

The writer is still engaged in the "Analysis of the World Oil Market" project through a joint agreement between the M.I.T. Energy Laboratory and the Center of Applied Research of the Norwegian School of Economics and Business Administration. The focus of the project is identical to that of this book, even if at a much more detailed level. The writer would not be doing so if he did not expect to see the problem differently and, he hopes, more deeply and to test ideas with other and more timely data. But the approach set forth here is, he ventures to hope, the correct one, and hence worth the attention of those interested in the subject.

Newtonville, Mass.
May 1976   Paul Leo Eckbo

 *Chapter One*

Introduction

THE INTERNATIONAL PETROLEUM MARKET

The price of crude oil in the international petroleum market, the imports-exports market, is determined in the same way as the price of any other product—by demand, production costs, and the kind and the extent of noncompetitive behavior. What makes the current price of crude oil interesting from a theoretic as well as an empirical point of view is the unprecedented degree of noncompetitive behavior that the current price level represents. The current price is fifty times the marginal production costs of the marginal source of petroleum, the Persian Gulf producers [1]. When the most important commodity in international trade becomes subject to such a degree of noncompetitive behavior, then the repercussions in the world economy for both consumers and producers of oil are many and complex. A formal framework to analyze the likely extent of noncompetitive behavior is, therefore, essential.

The circumstance faced in this market may be summarized in the following way:

1. There is a set of petroleum-importing countries, dominated by the industrialized economies of the U.S., Europe, and Japan. The net demand of each of these countries for imported oil is determined by the total energy demands of each of the countries, less the domestic supplies available, and less imports of other fuels, such as coal.

2. There is a fringe of petroleum exporters, which includes various non-OPEC sources, such as the producers of the North Sea, the

USSR, and China, and potentially other countries, such as Mexico. In this group may also be included some members of OPEC who have great needs for foreign exchange and who are "expanionist" in their oil-production policies, such as Indonesia, Iraq, and Nigeria.

3. A small group of Persian Gulf nations are the "price-makers." This group includes Saudi Arabia, Kuwait, and others in the Gulf; under some definitions it also may include Libya, Iran, and Venezuela. These countries face a *residual* demand for world oil, which is the total demand less that supplied by the fringe. The members of this group form the active core of the oil-exporters; their control over price involves two parts: the setting of the price itself, and the control of production so that it does not outstrip the residual world demand at that price.

The members of OPEC controlled 90 percent of the international market in 1973 and 73.3 percent of the world's proven recoverable reserves of oil (Table 1-1). Saudi Arabia alone controlled 23 percent of the imports-exports market and almost 25 percent of the world's proven recoverable reserves in 1973, the year the oil-exporters were able to increase the average cost of crude to the oil companies from a low of $1.62 to $9.25 (Table 2-1).

The structure of the international petroleum market is oligopolistic, and the exporters have been able to take advantage of the oligopolistic structure to raise the price of oil. An oligopolistic market cannot be analyzed with as much confidence as a competitive market. That some exporters have entered into an explicit agreement to limit competition for mutual benefit, a cartel agreement, does not reduce the uncertainty of the market solutions that emerge. We are faced with critical uncertainty regarding the ability of the Persian Gulf states to restrict production cooperatively, or the willingness of a single country to bear the burden of output restraint, so as to support the price received by all of the world's petroleum exporters. We also expect to observe deviations from competitive behavior by the noncolluding exporters, the exporter fringe. There is considerable uncertainty about the extent and the effect of various fiscal and nonfiscal regulatory regimes on the production rate in non-OPEC countries like Canada, Norway, and the United Kingdom. The future escalation of costs and the uncertainty with respect to the total number and location of geological traps containing hydrocarbons in commercial quantities, as well as the future path of international prices, render evaluations of the relationships between these policy parameters and the level of activity in non-OPEC countries extremely difficult.

Table 1-1. World Oil Production and Proved Reserves, 1973 (Estimated)

Producing Area	Million B/Day	Percent	Billion Barrels	Percent
Western Hemisphere	16,122	28.9	76.1	13.4
United States	9,225	16.5	34.6	6.1
Venezuela	3,370	6.0	14.2	2.5
Canada	1,750	3.1	9.7	1.7
Others	1,777	3.3	17.6	3.1
Western Europe	396	0.7	15.9	2.8
Middle East	21,337	38.3	350.3	61.7
Saudi Arabia	7,671	13.8	140.8	24.8
Iran	5,870	10.5	60.2	10.6
Kuwait	3,144	5.6	72.7	12.8
Iraq	1,960	3.5	31.2	5.5
Others	2,692	4.8	45.4	8.0
Africa	5,840	10.5	67.6	11.9
Libya	2,190	3.9	25.6	4.5
Nigeria	2,020	3.6	19.9	3.5
Algeria	1,020	1.8	7.4	1.3
Others	610	1.1	14.7	2.6
Asia-Pacific	2,275	4.1	15.9	2.8
Indonesia	1,330	2.4	10.8	1.9
Others	945	1.7	5.1	0.9
Communist Countries	9,780	17.5	42.0	7.4
USSR	8,400	15.1	34.6	6.1
China	1,000	1.8	7.4	
Others	380	0.7		1.3
World Total	55,750	100.0	567.8	100.0
OPEC Members[a]	30,837	55.3	416.3	73.3
OAPEC Members[a]	18,400	33.0	299.8	52.8

[a]The members of the Organization of Petroleum Exporting Countries in 1973 were Algeria, Ecuador, Indonesia, Iran, Iraq, Kuwait, Libya, Nigeria, Qatar, Saudi Arabia, United Arab Emirates, and Venezuela. The members of the Organization of Arab Petroleum Exporting Countries are Algeria, Bahrain, Iraq, Kuwait, Libya, Qatar, Saudi Arabia, and United Arab Emirates. Gabon was granted full membership in July of 1975.

Source: *Oil and Gas Journal*, December 31, 1973.

A distinctive peculiarity of a mineral industry is the need to take account of depletion. With any fixed stock of minerals, investment requirements and unit costs increase as depletion proceeds, both because of the tendency to go from more to less accessible deposits and because of the anticipation of higher future prices. On the other hand, rising prices may be retarded and even drastically reversed by technological improvement or discoveries of new deposits.

APPROACH OF THE STUDY

Given the structure of the international petroleum market and the critical uncertainties associated with the future development of this market, two broad areas of analysis were defined:

1. What is the likely future path of the residual world demand faced by the key cartel members?
2. What is the likely behavior of this group under those demand conditions, given what is known about the producers' reserves, production costs, resource needs, political goals, etc.?

The contractual relationships between the oil companies and the exporter countries reflect the degree of decision-making power that each of the two parties enjoys and the interdependencies of the two parties. To identify the location of decision-making power and the extent to which the possession of this decision-making power has been used or could be used to establish and/or to maintain a noncompetitive price level, a review was made of the contractual relationships as they have evolved over time. Chapter Two summarizes the findings of this review.

To learn about OPEC by analogy, a review was made of the experience of some previous international commodity cartels in Chapter Three. The experience of these cartels was coded to indicate central factors in the operation of workable versus unworkable cartel agreements.

There is also an extensive literature on oligopolistic markets and on cartels. This literature was searched for possible applications to the international petroleum market. Conventional oligopoly models do not, however, seem very useful when studying particular markets. Chapter Four concludes that a more detailed study, "story-telling," is needed to hypothesize which coalitions are likely to emerge under the different circumstances the exporters might face. The need to combine aspects of formal modelling with informal story-telling makes a simulation model the most ambitious approach that can be made to the analysis of a particular oligopolistic market without sacrificing the empirical validity of the analysis. The representation of the international petroleum market in the model reflects the need to combine formal and informal modelling, as reported in Chapter Five. The focus of this representation is on the behavioral characteristics of the exporters. A set of decision rules has been defined for each of the exporters, and an explicit analytic expression for the market price consistent with each combination of exporter decision rules has been derived. The evolution of combinations of decision rules to be observed over time, and the resulting price and quantity paths consistent with the evolution of these decision rules are determined by story-telling. Cartel theory, the experience of previous cartels, and the price responsiveness of the consumer markets as well as the special characteristics of the exporters as indicated in Chapter Six are used to construct the "stories." The "stories" are told and the results of the "story-telling" or of the model simulations are reported in Chapter Seven.

A simulation model has been constructed in response to the need for a formal framework to analyze the international petroleum market. The operation of such a model corresponds to the simultaneous solution of a set of time-dependent equations. The simulation model is highly adaptable to changed or improved functional inputs. That portion of the model that calculates demand for imports can be linked up with any one of several representations of exporter fringe behavior and of likely cartel behavior. The structure of the model is flexible enough to allow features of formal models to work with informal story-telling to explore the extent and implications of noncompetitive market behavior to be observed in the international petroleum market in the years to come.

STUDIES OF THE INTERNATIONAL PETROLEUM MARKET

The lack of formal models relevant to a study of a particular oligopolistic market is reflected in the approach taken in the major studies of the international petroleum market or of the world oil market. The studies made of this market are mostly informal studies of the story-telling or the industrial organization type. More recent attempts to introduce formal models have been made. To accommodate these models, however, the oligopolistic structure is assumed to be replaceable with either a perfectly competitive structure or a monopolistic structure.

The industrial organization study of a market allows the researcher to make snapshots of an industry from any angle. The richness of detail may be overwhelming, but if artfully sorted and put together in a consistent framework, an industrial organization study may give a better understanding of a particular noncompetitive market than any collection of formal models. An outstanding example of an artful study of the international petroleum market, the imports-exports market, in a larger world petroleum market context is

Adelman [1]. The underlying hypothesis of the study by Adelman, as well as the more recently published study by Jacoby [5], is that of a convergence towards a perfectly competitive market. These studies provide evidence that a number of markets that we include in the term "the world petroleum market," such as transportation and refining, were already competitive, and that the world petroleum market as a whole was actually converging against a competitive market structure in the 1960s. Adelman in his study, however, pointed out the unexploited market power of the producer countries and the instability that might result from this.

The evidence on an emerging competitive market and the need for quantitative rather than qualitative conclusions induced Houthakker and Kennedy [6, 8] to construct a formal model of a competitive structure. The inclusion of an exogenous export tax in their model as a proxy for the level of monopoly rent to be collected by a producing country does not alter the competitive structure of the model. By solving the model for any given future year, assuming an export tax level, the model provides projections of consumption, supply, and trade flows for that year. The Houthakker-Kennedy model is thus similar in kind to a number of other studies that have focused on the composition of the future world energy market under various assumptions about the price of crude oil in the Persian Gulf [4,9,11]. Neither OECD, Ford Foundation, nor FEA, however, consider the world market as a cartel problem. But these studies have made a significant contribution by evaluating some of the implications of OPEC pricing scenarios, and thus have indicated the responsiveness of the world's oil/energy markets to the price of crude oil in the Persian Gulf. A scaled-down version of the price responsiveness of the world's oil energy markets, as indicated by the studies of OECD and FEA, has been used as a basis in this study when determining some of the pricing strategies of the producer countries.

Allocation of crude oil and petroleum products was the problem approached by the large LP-type models traditionally formulated by the oil companies. The Deam effort [3] and the forthcoming Rapoport study [10] exemplify this tradition. The rigid structure of these models makes an analysis of the uncertainties we face today difficult.

Some recent efforts have been made to consider oil price setting from the producer countries' point of view. The model developed for the World Bank's Energy Task Force by Blitzer, Meeraus, and Stoutjesdijk simulates the world oil market under a set of OPEC pricing strategies and ranks the strategies according to some possible policy criterions [2]. Kalymon has developed two models to compute the long-term pricing strategies for OPEC which maximize the total discounted value of oil reserve exploitation for OPEC as a whole or for some subgroup, and to compute the desirability of various market-sharing mechanisms within OPEC to the membership of OPEC [7].

The World Bank model is informal in the sense that the world oil market is simply extrapolated under an arbitrary set of pricing strategies, and the outcomes of these pricing strategies are ranked according to how "likable" they appear. Kalymon has developed a model of a monopolistic structure which can be used to determine the "optimal" price path to the individual countries that are assumed to belong to the monopoly unit. The Kalymon approach is thus the extreme counterpart of the Houthakker-Kennedy approach, the former assuming a given and unchangeable monopoly structure, the latter a competitive structure.

In an oligopolistic market the price level may fall below the competitive level, as has happened in price wars, and also increase above the monopoly level, as in the case of a cartel exploiting its short-term market power and the uncertainty associated with the emerging price level. That is, if a cartel increases the price above a level considered to be the monopoly level, the market may not respond to the whole price increase because the market expects the price to come down, which implies that the cartel may benefit from an "uncertainty premium." That a competitive structure as well as a monopolistic structure might emerge from an oligopolistic market structure implies that a model designed to understand the likely development of such a market must be flexible enough to change structure or price behavior over time. The simulation model of this study is designed to explore the implications of pricing behavior reflecting different market structures in various periods over the simulation horizon. The model described here is evolutionary in the sense that each exporter is assumed to behave according to a set of decision rules that may reflect a competitive market structure, a monopolistic market structure, or any combination of the two. The change of the decision rules applied from period to period provides for discontinuities in the price path as we would expect in a cartel-dominated market. An attempt to combine the merits of formal competitive and monopoly models as well as the informal story-telling approach has been made.

 *Chapter Two*

Companies Versus Exporter Governments

The relationship between the oil companies and the governments of the exporting countries has changed dramatically over the last few years. A major aspect of this change is the transfer of decision-making power from the oil companies to the exporter governments. When analyzing a cartel-like organization like OPEC it is essential to determine to what extent the organization controls the instruments required successfully to operate a cartel. In the following review of the contractual relationships between oil companies and producer countries over time, the focus is on the location of the decision-making power and the extent to which the exploitation of this decision-making power has been used or could be used to establish and/or maintain a noncompetitive price level. The degree of decision-making power and the ability to use this power will determine the longevity of OPEC as a cartel-like entity.

THE PIONEERING PERIOD

When the oil companies began looking for oil in Latin America and the Far East, these areas were politically dominated by the industrial countries of the West. No local petroleum expertise existed. Petroleum was not considered an essential commodity. The industrial countries were therefore often in a position to determine both who should be given the rights to explore for and produce oil and the terms on which these activities were to be undertaken, the terms of the concession agreements.

The main characteristics of the earliest concession agreements (1901-51) were:

9

1. the large concession areas and the long duration of these concessions,
2. the small number of concessionaires,
3. the homogeneity and simplicity of the concession conditions,
4. royalties as the main financial compensation,
5. the modest financial compensation for the concession due to the low value of crude oil and the limited need for it in the earlier years as well as to the limited bargaining ability and bargaining power of the prospective producer countries, and
6. the slow development of the concession conditions.

The oil companies controlled completely all aspects of oil production and pricing. The governments received a royalty usually stipulated as a fixed nominal amount per ton lifted, and that was the extent of their involvement in the petroleum industry.

The companies were able to exploit their control of production and marketing. If we disregard the era of the Standard Oil monopoly, the first collusive agreement covering the international market was made in 1928 [4]. In 1928, Standard Oil of New Jersey and Shell agreed to maintain market shares in foreign markets and in the acquisition of foreign oil interests, the "as is" (or "Achnacarry") agreement. Jersey was a spokesman for all American exporting companies, which at that time made serious attempts to form a U.S. Webb-Pomerene Export Association, a device by which a group of U.S. firms may join their export activities without violating U.S. antitrust legislation. As far as prices were concerned, the high costs of the marginal U.S. wells, including many stripper wells, would determine prices everywhere in the world. Price was supposed to be equal to the U.S. Gulf price plus freight, regardless of the actual origin of the oil.

The collaboration between American companies within the Webb-Pomerene framework collapsed, however, in 1930 because of incomplete coverage of all sources of supply in each particular market. In the U.S. it was difficult to control output because of widespread ownership and antitrust legislation. In Europe, Russian and Rumanian producers caused problems. There was, however, no evidence of price warfare or retaliation following the collapse of the agreement.

In 1930 a new agreement was signed with detailed penalties for over- and under-trading. This agreement controlled the European markets even if Russian and Rumanian producers still caused problems. In 1932 Rumania entered into a tentative agreement. An attempt to incorporate the Russians failed.

The 1930 "as is" was declared to be applicable to all countries outside the U.S.. A "Draft of Principle" in 1934, of "Utmost Confidence," arming the arrangement with supply quotas, distribution quotas, and fines for over-trading, formally extended the coverage of the agreement to the whole world.

The uncontrolled producers were in the U.S., Mexico, Russia, and Rumania. The disruptive ability of this outside fringe was limited. As early as 1952 the U.S. was a net importer, Mexico was unimportant, and Russia and Rumania had disappeared behind the Iron Curtain. The noncompetitive price level previously established survived, even if no formal agreement was recorded beyond 1945[4].

Before World War II the U.S. Gulf was the only area in which there was to some extent free trade in crude oil. The notion of posted price originated from the fact that buyers of crude in the Gulf area publicly stated the prices at which they were willing to buy crude. The Gulf price, being a proxy for the open market price, was used as a basis for determining the price of crude elsewhere. The posted price of crude oil in the Middle East was determined by deducting transportation cost from the port of departure to the U.S. Gulf from the posted price in the U.S. Gulf. The posted price was also used as a tax-reference price. The companies paid royalties and taxes on the basis of the posted price, even if the price at which oil could be sold in the marketplace was below the posted price, which has been the case since the early 1950s.

THE SECOND GENERATION

The earlier concession agreements gave exclusive exploitation rights in enormous areas. These rights were given on terms that appeared unfavorable to the producer countries when they became aware of the market value of their petroleum. The agreements created monopolies for the concessionaires, with the accompanying animosity this provoked. The agreements also tended to transfer a country's oil policy-making from the legal governmental powers to the oil companies. The traditional concessions, held almost exclusively by the eight major international oil companies, were increasingly regarded as derogatory to the national honor. In a world of rising nationalism the concession agreements in their original form could not survive. In some countries the concession regime was simply terminated by nationalization of the oil industry: Mexico nationalized its oil industry in 1938; Iran followed in 1951 [8]. The petroleum expertise available in the producing countries was, however, very limited. A reasonable compromise between the emphasis on national sovereignty and the efficiency of the oil operations was therefore a revision of the existing concession agreements in favor of the producing country.

With the exception of Venezuela, no exporter country had developed petroleum expertise by the end of World War II, when the era of decolonization began and the third world got an opportunity to increase cooperation among themselves. Venezuela helped educate the producer countries by organizing an exchange of information and views on aspects of the oil industry and on oil policies. Venezuela also saw the opportunity for increased revenues if the producer countries could bargain with the oil companies on a joint rather than an individual basis. Venezuela promoted vigorously these ideas during the 1950s. The Arab League, formed in 1945 to provide the Arab oil-producing countries with a formal institutional structure within which to develop further collaborative arrangements, did not, however, mature to the point at which it could have realistically taken over the oil industry [2].

It was the entering oil companies, the independents and the state oil companies of the consumer countries, and not the producer countries themselves that eventually took the initiative in a revision of the earlier concession agreements. Venezuela was the originator of many of the new contractual concepts introduced. The newcomers had to offer the host governments better terms than the majors in the competition for access to crude oil supplies. The noncompetitive price level was still sufficiently high to accommodate both the higher share being paid to the producer countries and the independents' required rate of return on capital invested.

This second generation of concession agreements has the following characteristics:

1. The agreement is valid only for a well-defined limited area for a limited period of time.
2. The agreement contains rules about relinquishment after the expiration of certain periods of time.
3. The agreement is split into a reconnaissance and a production phase.
4. An income tax of 50 percent is levied.
5. There are higher, often graded, royalties. The calculation of royalties is made explicit in the agreement. The principle of royalty expensing is followed.
6. Other fees are clearly defined and fixed.
7. Rules for working programs and additional investments are fixed.
8. Detailed rules for the solving of disputes, prefixed rules about arbitrage, are formalized.

Table 2-2 indicates how royalty and income tax rates are used to produce a government take, the producer governments' per barrel revenues.

The fifty-fifty profit-sharing principle established in this second generation of concession agreements increased substantially the revenues of the producer countries. These agreements did not, however, give the producer countries access to the decision-making bodies. Pricing and lifting schedules were still decided by the companies.

The entrance of new firms, in addition to increasing the bargaining power of the exporter governments, also increased the level of competition in the marketplace. The entering firms put a downward pressure on the price of oil. After a ten-year period of nominal price stability, the difference between the posted price and the price at which oil could be sold in the market increased in 1960 to the extent that the companies unilaterally reduced the posted price, the tax reference price, and hence decreased the government take. The producing countries opposed the reduction, established OPEC, and fixed a posted price of $1.80, which was 5 cents above the previous level.

PARTICIPATION

Venezuela was still in the lead when the companies reduced prices in 1960. When the companies unilaterally reduced prices for the second time in August of 1960, Venezuela convinced the other oil-exporting countries, and the Organization of Petroleum Exporting Countries was formed in September 1960. The exporting countries had demonstrated that their technical knowledge and skills as well as their power position enabled them to oppose unilateral price reductions by the companies. The main objectives of OPEC were:

1. a stabilization of oil prices in the international crude oil market;
2. a coordination of oil policies in the member states;
3. a stable income to the producer countries, and an effective, economic, and continual flow of oil supplies to the consuming countries, and a just rate of return on the oil companies' invested capital.

OPEC managed to reverse the downward trend in nominal prices, but not in real terms. The real tax-paid cost of crude decreased by 28 percent in the 1960s (assuming a "normal" rate of inflation of 4 percent).

HUNT LIBRARY
CARNEGIE-MELLON UNIVERSITY

The entrance of new firms continued through the 1960s. In 1960, twenty-one independents were producing in the Middle East. In 1970, fifty-four independents and thirteen national oil companies in addition to the eight majors were producing in this area [3]. The "majors," however, still played the major role, even if decreasingly so. The share of the seven largest companies in non-U.S., non-Communist world oil production decreased from 87.1 percent in 1953 to 70.9 percent in 1972 [5].

The independents, with access to fewer sources of crude and financially weaker than the majors, were much more dependent on each producer country and thus also much more vulnerable to the actions of the individual producer countries. As the majors did not supply the independents with crude when the independents were cut off from their source by some producer country, individual producer countries were able to obtain considerable concessions from the independents. The concessions granted by the independents were later used as a basis for negotiations with the majors.

Again it was the vulnerability of the independents and the consumer-country-controlled state companies that made these companies, rather than the producer governments, introduce the contractual concepts that gave the governments direct control over domestic oil operations, and a means by which they could themselves participate in these operations. The Italian state oil company, ENI, finalized the first "joint-venture" agreements with government participation in 1957 with the state companies in Egypt and Iran [8].

The characteristics of the typical joint-venture agreements with government participation are:

1. The government's authority as a government and the government's rights as a participant in the venture are clearly separated.
2. The joint-venture is assigned a concession on the same terms as any other company. The government-controlled company thus becomes a concessionaire through the joint-venture.
3. The government-owned company is usually 100 percent owned by the government.
4. As a rule, participation is on a fifty-fifty basis.
5. The government-controlled company's risk is reduced through the principle of "carried interest." This means that the foreign company "carries" the interest of the government by assuming the entire economic risk until commercial discoveries are made.
6. When commercial discoveries are made, the government-controlled company generally has to pay in cash its part of the further costs associated with the future development of production facilities.

7. The government-controlled company takes out its share of the production in crude and may then sell the crude back to its partners or market the crude itself.

The government-owned oil companies operating nationalized industries did not have ready access to world markets. The Government Oil Refining Administration was set up in Iraq as early as 1952. The Iraqi National Oil Company was not formed until 1964 to engage in all aspects of the industry, including sales overseas, later taking over exploration and production under a nationalization program. In the years following the nationalization, the Iraqi oil industry stagnated, and it was not revived until 1973, when an agreement including compensation was reached with the majors [8]. Iran had a similar experience following the nationalization of the Iranian oil industry in 1951, a takeover that did not become a reality until 1973, when a long-term supply agreement was reached with the majors.

The lack of technical and managerial expertise in the exporting countries proved costly in terms of reduced oil activities. The need for assistance from the concessionary companies was therefore recognized. An effort was made to find a working relationship that could meet the requirements of the private companies without involving an equity participation. Indonesia, after having nationalized the oil industry in 1960, introduced production-sharing contracts in 1966. A production-sharing contract entitles the companies to a fixed percentage of the crude oil produced to recover their exploration and production costs. The formal ownership of the reserves discovered is retained by the state company. A politically explosive issue is thereby avoided. Politically neutral is also the "service contract" or "entrepreneur contract" concept formulated by Venezuela in 1959 and implemented for the first time in 1966 in an agreement between the French state-owned company ERAP and Iran. Under such an agreement the foreign oil company operates as a contractor for the state-owned company. The discoveries made and the petroleum produced remain the property of the state. The foreign oil company, the contractor, will usually get no more than a long-term right to buy a preassigned part of the produced crude at a discounted price. ERAP's agreement with Iran and Iraq in 1968 gave the company a long-term right to buy at a discounted price. However, in the much more significant agreement in 1973 between the major consortium companies in Iran and the National Iranian Oil Company, NIOC, no discounts were granted.

Some of the major Persian Gulf producers (Saudi Arabia, Abu

Dhabi, and Qatar), rather than announce nationalization of the oil industry, purchased "at a bargain price" a 25 percent share in the concessionary companies in an agreement of January 1, 1973 [8]. The companies were compensated on the basis of the book value of their assets only. Because none of the state companies was yet in a position to market oil on this scale, the governments required the oil companies to buy back the oil the governments obtained through the participation agreements at a price to be determined by the governments. The buy-back price has been considerably higher than the total tax-paid cost of equity oil.

Since 1973 the governments' share of the production ventures has increased rapidly. Kuwait recently announced a 100 percent "participation," in effect a complete nationalization of the industry [9]. The other producers in the Gulf and Nigeria are expected to follow the example soon. When the governments push for a 100 percent interest in the operating companies, the price structure is likely to converge towards the buy-back price, which would then become the price of crude.

It is still somewhat unclear how the companies will be compensated in the future, whether through a discount on the price of crude, a fee to perform under a service or a management contract, or some production-sharing arrangement. The producer countries will still need the services of the international oil companies for a number of years.

PRICING

The fast-rising demand for imported oil in the latter half of the 1960s put a strain on the delivery system. The spot tanker rate increased steadily from 1967. In May of 1970 the trans-Arabian pipeline (Tapline) was blocked by Syria [1]. The producing governments in North Africa exploited their favorable location by demanding higher taxes. When the companies refused to pay the higher tax, Libya restricted output to force the companies to agree to its demands. The companies rushed to the spot tanker market to secure transportation from the Persian Gulf. The supply of tanker transport is very inelastic in the short run when the industry is already operating at capacity. The spot market rate consequently jumped dramatically, and the companies found it economically justified to agree to the Libyan tax increase.

President Gadaffi of Libya demonstrated the ability and the power of an individual producer country to dictate its own terms. His initial success and the continued strain on the delivery system encouraged

President Gadaffi to push for better terms, and he emerged as a price leader of the producer countries.

The companies tried to recapture their bargaining power by agreeing upon a strategy of industrywide bargaining. They figured that the weakest countries' need for a settlement would discipline OPEC as a whole. The companies wanted to put pressure on OPEC's "independents."

The exporting countries, however, opposed such industrywide bargaining. They wanted first an agreement for the Persian Gulf and then, with this agreement as a precedent, to negotiate agreements for the other producing areas. The Persian Gulf was, in 1970 as today, the marginal source of petroleum in the world, and the producer countries of the Persian Gulf were therefore in the strongest bargaining position.

The issue of industrywide, as opposed to regional, negotiations brought company executives and OPEC officials to Teheran in February 1971. An agreement seemed impossible, when OPEC launched a very risky venture. On February 3 the OPEC representatives presented the companies with an ultimatum. If no agreement was reached by February 15, the separate governments would enforce their terms by legislation, imposing a ban on shipments by any company that refused to conform [7]. The companies elected to meet the OPEC demands, which implied an immediate increase in the governments' take of about 33 cents and acceptance of four additional phased increases in the posted price, on June 1, 1971, and on January 1, 1973, 1974, and 1975. With an agreement established for the Persian Gulf, agreements for Mediterranean and African crudes were reached in Tripoli in April.

The Teheran and Tripoli agreements constituted the first significant evidence that a cartel was organized and operating among the petroleum-exporting countries of the world. Since then the OPEC countries have been trying to climb up along the demand curve to benefit from a monopoly or a joint-profit-maximizing price.

The October 1973 cutback of production and the following increase in prices may, however, be explained by a political as well as by an economic hypothesis. As the October 1973 events certainly do not contradict an economic hypothesis, it was decided to focus primarily on economic behavior in this study.

Whether intentionally or not, from an economic point of view the timing of the October 1973 cutback was excellent. Demand for petroleum imports was growing close to the 1962-73 average of 7.5 percent per year. Partly owing to production and price controls of natural gas and crude oil in the United States, large increases in imports of crude oil and petroleum products were required. For market clearing at the regulated (frozen) prices, imports would have had to increase by 2 million barrels per day in 1973 [6]. The U.S. was thus on a 20 percent annual increase in imports path in 1972-73, very vulnerable to cutbacks, and thus easily convinced that higher prices were necessary.

When it was clear that the U.S. would not be able to respond to higher prices, at least not in the short run, no major consuming country was able to respond. Neither Western Europe nor Japan have short-term alternatives to their petroleum imports. There was consequently room for a major increase in the price of crude.

The escalation of the cost of crude to the oil companies and of the level of government take-over the last fifteen years is summarized in Table 2-1. The buy-back price, the buy-back percentage, and hence also the average cost of crude for the first six months of 1974 was determined retroactively. The companies were producing oil in that period without a clear idea of what they would have to pay for it. The intentions of the producing countries seem to have been to do away with the posted price system. The producer governments expected to sell their oil in the open market at a price that consisted of the $7 floor and a company margin that was not well defined. It soon appeared, however, that the governments could not sell anything like their buy-back amounts, and hence forced the companies to take them. The ability of the companies to take the buy-back oil and resell it at a very high price showed the governments that they could raise the price. The producer countries thus went through an important learning process in the spring of 1974. OPEC demonstrated an impressive level of discipline in the same period, being able to increase the price significantly at a time when excess capacity also increased significantly.

The price of crude oil in the Persian Gulf, the base point, and a set of quality and transportation differentials determine the price other producers may charge for their crude. The way the producer countries determine what the companies shall pay for crude oil and the dramatic change in their terms of production are demonstrated in Table 2-2. The posted price, or the tax reference price, which is now being unilaterally determined by the producer countries, serves as a basis for determining the producer countries' per barrel revenues. The royalties are calculated as a percentage of the posted price and are being expensed for income tax purposes. A per barrel income figure is determined by subtracting production costs and royalty payments from the posted price. This income figure is then used as a basis for calculating a per barrel income tax, which today is 85

Table 2-1. The Cost of Crude in the Persian Gulf
(Arabian Medium, 31 API, Ex. Ras Tanura)

Period	Posted Price	Government Take	Tax-Paid Cost	Buy-Back Price	Percent of Total Bought Back	Average Cost of Crude
1960-1965	1.80	0.82	0.92			
1966-1967	1.80	0.85	0.95			
1968-1969	1.80	0.88	0.98			
1/1-11/14, 1970	1.80	0.91	1.01			
11/15/70-2/14/71	1.80	0.99	1.10			
2/15-5/31/71	2.18	1.26	1.37			
6/1/71-1/19/72	2.28	1.32	1.43			
1/20/72-1/1/73	2.48	1.44	1.55			
1/1-3/31/73	2.59	1.51	1.62			
4/1-5/31/63	2.75	1.61	1.71			
June 1973	2.90	1.70	1.80			
June 1973	2.95	1.74	1.84			
August 1973	3.07	1.80	1.90			
1/1-10/15/73	3.01	1.77	1.87	2.80	20	2.05
10/16-12/31/73	5.12	3.05	3.15	4.76	20	3.47
1/1-3/1/74	11.65	7.00	7.10	7.10	25	7.10
3/1-7/1/74	11.65	7.00	7.10	10.82	25	8.04
1/1-7/1/74	11.65	7.00	7.10	10.87	57	9.25
7/1-10/1/74	11.65	7.06	7.16	10.96	62	9.50
10/1-11/1/74	11.65	8.12	8.22	10.96	47	9.81
11/1/74-1/1/75	11.25	9.82	9.92	10.67	58	10.35
1/1/75-	11.25	9.82	9.92	10.46	56	10.22

Sources: *Business Week* and *Petroleum Economist* and *Financial Times*.
N.B. Prices during 1974 under current conditions, not retroactively.

percent of this imputed income figure. The government's per barrel revenues on equity oil, that is, on the oil that the oil companies own according to the concession agreements, are hence equal to the sum of the per barrel royalty and income tax. The cost of buy-back oil and the percentage of the total being bought back by the companies from the governments participating in operating ventures are additional instruments the producer countries can use to make the weighted cost of crude, that is the average per barrel cost of crude to the oil companies, equal to some target price. The tax legislation of the consumer countries makes it favorable for the oil companies to

Table 2-2. Calculation of the F.O.B. Cost of Crude in the Persian Gulf

	January 1971	November 1974
Posted Price	$1.800	$11.251
Less Production Cost	0.10	0.10
Less Royalty	0.225	2.250
Taxable Base	1.475	8.901
Income Tax	0.738	7.566
Royalty	0.225	2.250
Government Take (Equity Oil)	0.963	9.816
Production Cost	0.10	0.10
Total Tax-Paid Cost (Equity)	1.063	9.916
Buy-Back Price	–	10.671
Companies' Average Cost:	–	
Equity Oil	1.063	9.916
Buy-Back Oil	–	10.671
Weighted Average	1.063	10.353
Typical Company Margin	0.35	0.35
F.O.B. Cost of Crude	1.413	10.703

January 1971:   Royalty 12.5 percent; Tax 50 percent
November 1974:   Royalty 20 percent, Tax 85 percent, buy-back price 94.85 percent of
posted price, 42.1 percent equity oil, and 57.9 percent buy-back oil.

realize the profits of their integrated operations in the producer countries, as the companies get a tax credit for the taxes being paid to the producer countries, and thus can partially eliminate tax liabilities elsewhere. The company margin used to calculate the transfer price of crude oil, or the F.O.B. cost of crude in the Persian Gulf is thus a proxy for the companies' profit margin on their integrated operations.

It is likely that the tax rules applicable to oil companies will be changed so that profits will be distributed more evenly on their downstream operations by removing the tax credits as they are formulated today. The emerging working relationship between the companies and the governments may imply that the rather arbitrary determination of the cost of crude (Table 2-2), resulting from the myriad of contractual and concessionary agreements presently in operation, will be replaced by some single price concept.

SUMMARY AND CONCLUSIONS

The "majors" completely dominated the international petroleum market up until 1950. The early concession agreements kept the involvement of the producer governments at an absolute minimum. The companies were able to exploit their market power by establishing and maintaining a noncompetitive price level. The early concession agreements were increasingly considered offensive to the national pride and could not survive in a world of growing nationalism. The noncompetitive price level induced smaller independent oil companies to establish themselves as integrated companies with access to equity crude. Nationalism also induced consumer countries to establish their own state oil companies to explore for and produce oil. The entering firms were able to obtain concessions by offering better terms to the producer countries. A second generation of concession agreements was based on the fifty-fifty profit-sharing principle introduced by Venezuela. Venezuela was the only producing country that played an active role in the conceptualization and implementation of these concession agreements, which helped all the producer countries reach their goal of higher revenues from the petroleum sector. The continued entrance of new firms during the 1950s and the 1960s and the vulnerability of the independents to the actions of individual producer countries induced the independents to introduce the concepts that would make it possible for the producer countries to reach their second goal, direct access to and control of the operating units. The 100 percent "participation" agreements that have been introduced lately, which mean in effect that the industry has been nationalized, give the producer countries a control constrained only by their level of technical and managerial expertise.

The development of the contractual relationships is summarized in Table 2-3, where the two major characteristics of the contractual concepts in their "pure" form are listed, as well as their dates of

Table 2-3. Characteristics of Contractual Concepts

	Date of Introduction	Percent Government Ownership	Government Percent of Crude Oil "Profits"
Concessions			
First Generation	1880	0	Nominal
Second Generation	1948	0	50[a]
Nationalization	1938	100	60-100
Joint-Venture Participation	1957	50	75
Service Contracts	1966	100	75-100

[a]58 percent with royalty expensing, which was introduced in the Middle East in 1964.
Sources: *Petroleum Economist*, Neil H. Jacoby, *Multinational Oil*, pp. 108-10.

introduction. Except for the first generation of concession agreements, all the contractual concepts listed in Table 2-3 are included in some form or another in the existing contractual relationships between the producer countries and the multinational oil companies.

The downward pressure on price resulting from the increased competition from the independents made the companies unilaterally decrease the posted price, the tax reference price, in 1960. Venezuela seized the opportunity to convince the producing countries of the need for an organizational counterpart to the oil companies, and OPEC was formed in September 1960. While OPEC managed to reverse the downward trend in nominal prices, but not in real terms, in the 1960s, it seems to have had no significant impact on the market price of crude oil.

The tanker shortage that developed in 1970 and the resulting dramatic increases in the spot market rates made it possible for President Gadaffi of Libya, which is located closer to the consumer markets, to obtain more for Libyan crude. The continued strain on the delivery system and President Gadaffi's resulting continued success in convincing the companies to pay higher taxes made him an ideological as well as a price leader of the exporters. The cohesiveness and the confidence of the exporting countries increased dramatically following Gadaffi's success. In Teheran and Tripoli in 1971, OPEC was able to force its demands through as a cartel. The exporter countries had decided to collude to enforce their demands. The strength of the colluding group was not tested, for the oil companies chose to avoid a confrontation by agreeing to the increase in government take.

The October 1973 cutback of production demonstrated the cohesiveness of the dominating Arab subset of OPEC. OPEC has since then also shown an impressive level of discipline by being able to live with a considerable excess capacity without cutting prices. President Gadaffi seems to have lost his leading role. Saudi Arabia, Egypt, and Algeria tend to dominate the political stage, with Algeria and Iran as price-leaders. Two years after the embargo, political pragmatism and economic development seem to dominate the decisions of the producing countries in the Middle East. This does not imply, however, that the relationship between the producer countries, especially in the Persian Gulf, does not have a major political ingredient affecting their oil policies; neither does it imply that one or more major producers will not again use their control over oil, volumes and prices, to gain political favors from the West.

The presently emerging contractual regime is giving the individual exporter governments control over production of oil. OPEC as an organization has, however, no instruments available to regulate prices and production rates. Production programs, or pro-rationing, has been discussed but never implemented. OPEC is no more than a forum for discussions and for coordination of the price and production policies of the individual exporting countries. Except for the threat of cutoff in Teheran and the October 1973 embargo, the producer countries have never made joint decisions affecting the rate of production. Even during the embargo period the production policies of all the OPEC countries were not perfectly coordinated. Iran, the most significant nonconformist, increased production significantly, while the Arab subset of OPEC restricted output [11]. Since the first major tax or price increase, the producer countries have accepted the way the marketplace and the lifting schedules of the oil companies have allocated production and profits among the exporters. The producer countries have thereby been able to avoid confrontations over these critical issues.

The recent accomplishments of the exporting countries cannot be explained by their experience in operating an oil industry. The experience of the countries that nationalized the oil industry (Algeria, Indonesia, Iran, Iraq, and Libya) is somewhat more advanced than the experience of the countries that followed a participation-path (Abu Dhabi, Kuwait, Nigeria, Qatar and Saudi Arabia). Venezuela, which has also decided to nationalize the oil industry, is the most experienced with respect to making decisions on prices and production rates.

Even if the oil companies have demonstrated that it is possible to maintain a noncompetitive price level in the petroleum industry as long as the colluding agreement covers a sufficient share of total supplies, the lack of experience in coordinating production rates, the lack of access to the final consumer markets, the heterogeneity of the producer countries, and the increasing government involvement in the marketing of crude, leading possibly to confrontations the exporters are not prepared to face, may make it difficult or impossible for the producer countries to agree on a common tax or price policy. To determine the extent to which these factors have affected the success and the longevity of cartel-like collusive arrangements, a review was made of the research findings on previous cartels reported in Chapter Three. The success of the exporters in sticking to an increasing price in the light of the level of excess capacity has, however, been so impressive that they may be reluctant to change the present market- and company-determined system of allocating production and profits. This system may be considered fair and just as long as there is no evidence that the companies deliberately try to divert the way the marketplace ranks the competitiveness of the various countries' crudes.

The oil companies are not yet powerless. They provide a service for which there is no institutionalized organizational substitute. In the foreseeable future, the oil companies will still be indispensable to an orderly growth of production and distribution in the international oil industry. This is true not only because of their technical and logistics capacity, but more important, because the companies constitute total systems wherein all of the various parts are integrated and attuned to one another. The significance of their logistics capacity was demonstrated during the embargo, when the companies were able to partially divert supplies, thereby reducing the effect on the primary targets of the embargo, the Netherlands and the U.S. [11].

The oil companies will probably continue to be the target of both producer and consumer dissatisfaction. From the perspective of the producer countries, collusion between the oil companies and the consumer countries seem to be a fact. The companies did reduce prices in 1960 and in real terms also in the 1960s. During the last year the oil companies have been accused by the consumer countries of colluding with the producer governments. To the extent that any company owns petroleum reserves, such a company shares the interests of OPEC in high world market prices for crude oil.

The companies are severely constrained, however, in their ability to resist producer government initiatives, even if the oil companies are an important component of the mechanism for raising and supporting international oil prices.

※ *Chapter Three*

The Experience of Previous International Commodity Cartels

INTRODUCTION AND SUMMARY

The Teheran and Tripoli agreements of 1971 constituted the first significant evidence that a cartel was being operated by the oil-exporting countries. The price-making power of OPEC was clearly demonstrated in the fall of 1973 and in 1974. The average price of crude in the Persian Gulf increased by 505 percent between October 15, 1973 and November 1, 1974. As long as the OPEC countries can agree on a joint market strategy, they can take advantage of their monopoly power and enjoy monopoly profits. A cartel is, however, an unstable unit even from a theoretical point of view. The market solution resulting from explicit collusion among oligopolists cannot be uniquely determined. The OPEC countries are also a rather heterogenous group of countries.

The economics and political science literature contains more than fifty studies of the operation of cartels in the trade of international commodities. Agreements have been formed by companies and countries in commodities as far-ranging as tin and tea; these agreements have lasted for varying lengths of time with varying degrees of success in curtailing production and raising prices to consuming countries. The research literature has documented cartel "success" and has provided a number of reasons why some cartels have worked better than others.

Although there have been numerous cartels in almost every commodity in international trade, only a small number of these price-controlling organizations have been studied in detail. Of those studied, even a smaller number have been analyzed completely enough to make it possible to tell the difference between cartel success or failure. We have found evidence on fifty-one cartel agreements in eighteen industries. These constitute two samples from which we draw conclusions on the factors determining the success or failure of cartels.

Cartel success or "efficiency" has been defined in terms of the ability of the organization to raise price at least 200 percent above the unit cost of production and distribution. If the cost to the highest cost member of the cartel at the margin were $1 per ton, then the cartel would be efficient if it raised prices to $3 per ton and kept them there for a significant period of time.

The review indicates that of the fifty-one significant cartel organizations reported, only nineteen achieved price controls that raised the level of charges to consumers significantly above what they would have been in the absence of agreements. But even the efficient cartels did not seem to last very long. Cartels were able to raise prices for four years or more, where *concentration* of production was high, *demands inelastic*, the cartel's *market share* was high, the membership had *cost advantages* over outsiders, and governments did not get involved in the operations of the cartel. In the industries where the structure was most favorable for efficient collusive arrangements, attempts to reestablish a collusive agreement, once an agreement had fallen apart, were often successful.

If OPEC were to follow the pattern set by the nineteen earlier "efficient" cartels, then it would likely have a four-to-six year duration. The primary source of breakdown would likely be the uncontrolled additions of supply from the "fringe" of OPEC countries (Iraq, Indonesia, Nigeria) or from the non member countries. Even if OPEC should fall apart, it is unlikely that no further attempts to form a collusive arrangement among the major oil-exporters would be made. In fact, it is likely that some of the post-OPEC attempts to form an international petroleum cartel would be successful.

CARTEL CHARACTERISTICS

The information available on international cartel agreements is not sufficient for rigorous empirical hypothesis testing. The studies made on cartels differ significantly in terms of their level of detail and research focus. From a theoretical point of view all important aspects of a cartel agreement were not covered in the studies reviewed. We therefore constructed two samples out of the fifty-one cartel agreements on which we had enough information to judge whether the

The Experience of Previous International Commodity Cartels 27

cartel agreement had been successful or not. To identify the most important factors determining the "efficiency" and longevity of cartel agreements, each cartel was summarized along a number of dimensions. Owing to the anecdotal and/or vague nature of the data, we have been limited to a very tight range of response, often to binary representation. On the cartels belonging to sample 1 we had sufficient data to describe the cartels along seventeen dimensions. Sample 2 consists of cartels on which we had sufficient data to code five dimensions only. The dimensions are intended to describe as completely as possible the known occurrences of cartel formation. The unavailability of information on the internal operating mechanisms of these cartels made it impossible to include these important aspects of a cartel agreement.

Cartel Characteristics of Sample 1

Dimension 1. The concentration of production in the industry is regarded as being high if the four largest producers in the industry produce more than 50 percent of the total output of the industry. If this is the case, the industry gets a score of 1; otherwise, a score of 0 is assigned.

Dimension 2. The concentration of the international market, the exports-imports market, is scored in the same way; if the four largest exporters constitute more than 50 percent of the total market, the score of this dimension is 1; otherwise it is 0.

Dimension 3. The elasticity of demand is also scored in a binary way. If the elasticity of demand is more than 1, the score is 1. If the elasticity of demand is less than 1, that is if the demand is relatively inelastic, the score is 0. As the decision horizon of the cartel usually seems to be shorter than the period needed to get a long-term adjustment to prices, it is the short-term elasticities that are considered relevant.

Dimension 4. The income elasticity is given a value of 1 if demand for the commodity is income-elastic, that is if a percentage change in income implies an even larger percentage change in the demand for the commodity; otherwise, the value of 0 is assigned to this dimension.

Dimension 5. If short-term substitutes for a commodity exist, the value of 1 is assigned; if no substitutes exist, 0 is assigned to the commodity.

Dimension 6. The existence of long-term substitutes is treated the same as for dimension 5.

Dimension 7. If governments were involved in the cartel agreement, a value of 1 is assigned; otherwise a value of 0 is assigned.

Dimension 8. The length of survival of the formal agreement in years.

Dimension 9. If the cartel members' share of total production in the industry is above 75 percent, a score of 2 is assigned; if the cartel members' share is between 50 and 75 percent a score of 1 is assigned. A score of 0 is assigned if the share is below 50 percent.

Dimension 10. If the cartel members are responsible for more than 75 percent of total exports in the international export-import market, a score of 2 is assigned. A score of 1 indicates that the cartel members export between 50 and 75 percent of the total; a value of 0 is assigned if the cartel members export less than 50 percent of that particular commodity.

Dimension 11. This dimension is included to test whether industries learn over time, that is, if the number of previous attempts to organize a cartel influence the success of later attempts to organize. The score is equal to the particular cartel's number in this sequence of attempts.

Dimension 12. Members are given a score of 1 if the cost differences within the cartel are less than 50 percent, that is, if the high-cost producers produce at a cost no larger than 50 percent above the low-cost producers. Otherwise, the dimension is given a score of 0.

Dimension 13. The efficiency of the cartel refers to the ability to charge prices close to the monopoly price, i.e., if price is 200 percent of marginal cost or more. Otherwise, the score of 0 is assigned. This very rough indicator of cartel efficiency was applied because information on the location and slope of the demand curve and the location and slope of the marginal cost curve usually was not sufficient to allow calculation of the monopoly price.

Dimension 14. This dimension is given a score of 0 if the cartel members' potential time horizon is more than one year and a score of 1 if the time horizon is less than one year.

Dimension 15. The dimension is given the score of 0 if a cartel breakdown was not market-related, i.e., due to government intervention, war, etc., and a score of 1 if the breakdown was market-related, i.e., due to the loss of markets to outsiders or the emergence of competition between cartel members.

Dimension 16. This expands on dimension 15 by assigning a value of 1 if the breakdown was market-related and due to external forces, i.e., nonmember suppliers or consumer retaliation, and a value of 0 if the cartel broke down due to an internal conflict between the cartel members.

Dimension 17. This final measure further expands on the breakdown issue by assigning a value of 1 if the external forces were outside supply, i.e., nonmember suppliers, and a value of 0 if the response of consumers or demand response constituted the external forces that caused the cartel breakdown.

Cartel Characteristics of Sample 2

This second sample was necessary because we did not have sufficient information to characterize the cartels along the full set of seventeen dimensions shown above. The attributes of sample 2, therefore, should be viewed as a quick summary and are essentially a subset of the attributes of sample 1.

Dimension 1. This refers to the length in years of the agreement. (See dimension 8 above.)

Dimension 2. This attribute is similar but not identical to dimension 1 of sample 1. If the four-firm concentration ratio is more than 75 percent, a score of 2 is assigned; a 1 is given if between 50 and 75 percent. Concentration of less than 50 percent is designated as 0.

Dimension 3. Here we are referring to concentration within the cartel itself. (See dimension 9 of sample 1.)

Dimension 4. Cartel breakdown is analyzed as in dimension 16 in sample 1.

Dimension 5. Cartel efficiency is described as in dimension 13 in sample 1.

SAMPLE 1

Sample 1 consists of the industries on which we were able to obtain information to assign a numerical value to the seventeen dimensions defined above. Ervin Hexner's *International Cartels* and G.W. Stocking and M.W. Watkins' *Cartels in Action* are the basic sources of information. In addition, however, it has been necessary to apply information given by Dr. James C. Burrows in his testimony before the Subcommittee on Economic Growth, July 22, 1974, some recent articles on international commodity markets, as well as our personal judgment. Because of the rather superficial scanning of the existing cartel literature, as well as the rather inaccurate state of the data given in this literature, a critical attitude on the part of the reader is recommended. By going through the cartel experience of various industries, and also explaining the way that we have coded this experience, we hope to give a feeling for the difficulties involved when trying to characterize cartels on the basis of such information. The results are shown in Tables 3-1 and 3-2.

Natural Rubber

The Stephenson Plan, which was sponsored by the British government, was an attempt to regulate the rubber industry. Even though the plan was a short-term success, it later failed completely. The plan lasted from 1922 to 1928. The large number of relatively small plantations made the rubber industry and the rubber trade fairly decentralized. According to Stocking and Watkins, demand for rubber was inelastic at that time. There were no substitutes for rubber in the production of tires and tubes. Synthetic rubber was, however, in the process of development.

The British colonies contained 72 percent of world capacity in 1922. The Dutch colonies contained another 25 percent of world capacity. The Dutch twice refused cooperation, but took advantage of the plan by increasing production. The British market share decreased from 67.5 percent in 1922 to 54.1 percent in 1927, whereas the market share of the Dutch colonies increased from 23.2 percent to 37.7 percent.

Outside production, internal rivalry, as well as the problems of timing of restrictions were the reasons for the failure of the Stephenson Plan.

The International Rubber Regulation Agreement of 1934 did, however, succeed in increasing prices so that the average producer, according to Stocking and Watkins, could enjoy a margin of 126 percent, and we judged the cartel to have been efficient, given the fact that the cartel lived with the threat of synthetic rubber.

Table 3-1. Characteristics of Efficient Cartels, Sample 1

	Rubber	Mercury		Aluminum					
Year	1934	1928	1939	1901	1906	1912	1923	1929	1931
1. Concentration of Production	0	1	1	1	1	1	1	1	1
2. Concentration of Exports/ Imports	0	1	1	1	1	1	1	1	1
3. Demand Elasticity	0	0	0	0	0	0	0	1	1
4. Income Elasticity	1	1	1	1	1	1	1	1	1
6. Long-term Substitutes	1	0	0	1	1	1	1	1	1
7. Government Involvement	1	0	0	0	0	0	0	0	0
8. Length of Formal Agreement	6	8	10	5	2	2	3	4	5
9. Cartel Members' Share of Total Production	2	2	2	0	0	0	0	0	2
10. Cartel Members' Share of Exports/Imports	2	2	2	2	2	2	2	2	2
11. Number of Recorded Attempts to Set Prices	4	1	2	1	2	3	4	5	6
12. Cost Differences Among Cartel Members	0	1	1	1	1	1	1	1	1
13. Efficiency	1	1	1	1	1	1	1	1	1
14. Potential Time Horizon of Agreement	0	0	0	0	0	0	0	0	0
15. Breakdown Market Related	0	0	1	1	1	0	0	0	0
16. Breakdown Externally	—	—	0	0	1	0	1	1	1
17. Breakdown Due to External Supply	—	—	—	—	1	—	—	1	0

Table 3-2.   Characteristics of Inefficient Cartels, Sample 1

Year	Rubber 1922	Tin 1929	Tin 1931	Tin 1935	Sugar 1931	Sugar 1937	Sugar 1958	Steel 1926	Steel 1930	Steel 1931	Steel 1933	Tea 1933	Copper 1955	Copper 1964
1. Concentration of Production	0	1	1	1	0	0	0	0	0	0	0	0	1	1
2. Concentration of Exports/Imports	1	1	1	1	1	1	0	0	0	1	0	0	1	0
3. Demand Elasticity	1	1	1	1	1	1	0	1	1	1	1	1	1	0
4. Income Elasticity	1	0	0	0	1	1	0	0	1	0	1	1	0	0
5. Short-Term Substitutes	1	0	0	0	1	1	0	0	0	0	0	1	1	1
6. Long-Term Substitutes	0	0	0	0	1	1	1	0	0	0	0	1	1	1
7. Government Involvement	1	1	1	1	0	1	1	0	0	0	0	0	1	1
8. Length of Formal Agreement	6	2	3	2	4	2	3	4	0	0	6	6	2	2
9. Cartel Members' Share of Total Production	1	2	2	2	0	2	2	0	0.5	0.17	2	1	0	2
10. Cartel Members' Share of Exports/Imports	1	2	2	2	2	2	2	1	1	1	2	2	0	2
11. Number of Recorded Attempts to Set Prices	3	1	2	3	5	6	?	?	?	?	?	?	?	?
12. Cost Differences Among Cartel Members	0	0	0	0	0	0	1	1	1	1	1	1	1	1
13. Efficiency	0	0	0	0	0	0	0	0	0	0	0	0	0	0
14. Potential Time Horizon of Agreement	0	0	0	0	0	0	1	0	0	0	0	0	0	0
15. Breakdown Market Related	1	1	1	1	0	0	0	1	1	0	0	0	0	1
16. Breakdown Externally	1	0	0	0	1	0	—	1	0	0	—	—	1	1
17. Breakdown Due to External Supply	1	—	—	—	—	—	—	—	—	—	—	—	0	0

The British, Dutch, French, Indians, and Siamese kept the agreement up until World War II, even though attacked by U.S. protests, which resulted in the organization of a semiofficial resistance movement to conserve tires and use reclaimed rubber.

The consumption of rubber was assumed to be income-elastic.

Tin

Production of tin was dominated by a few governments in the Far East—Malaysia (Dutch), Thailand, Nigeria, and the Belgian Congo. These producers tried to regulate tin prices, but recorded attempts in 1929-31, 1931-35, and 1935-37 were all failures owing to lack of discipline and enforcement of the restrictive measures.

There was no satisfactory substitute for tin, even though there was some secondary recovery from scrap. Tin was indispensable in armaments, and we assumed that demand was inelastic, as is also the case today, according to C. Fred Bergsten. The statement by Hexner that "production costs varied from mine to mine" is the basis for our assumption that costs differed by more than 50 percent.

Mercury

According to Hexner and Burrows, the price of mercury has been close to the monopoly price since 1928. Spain and Italy have completely dominated the production of this commodity, for which no substitute exists. As mercury is also indispensable in armaments, price-inelastic and income-elastic demand is assumed.

The cost difference between Spanish and Mexican producers is assumed to be above 50 percent. The cartel established in 1928 broke down in 1936 because of the Spanish War. It was reestablished in 1939 and then lasted until 1949, when it broke down owing to internal problems. Since 1950 there have never been more than three years of disagreement among the major mercury producers of the world.

Aluminum

Originally because of patent rights, and later intercorporate ties, the aluminum industry has been highly concentrated. The sequence of cartels (1901-06, 1906-08, 1912-14, 1923-26, 1926-30, and 1931-36), all seem to have been successful in stabilizing the monopoly level of the previous period.

According to Donald H. Wallace, the elasticity of demand increased in the late twenties following the conversion of latent into effective demand through the development of new alloys and products. Aluminum became at this time a capable substitute for various alloys of iron, copper, and zinc in heavy-duty components. The aluminum industry was undergoing a transition from a condition of limited markets to one of diversified markets. We therefore assumed that demand moved from the inelastic to the elastic segment of the demand curve in the late twenties.

Demand also seems to have been income-elastic in this period. The importance of technology should imply that cost differences were small. The capital-intensity of consumption seems to indicate that no short-term substitutes existed even if long-term substitutes did exist.

Steel

The first international steel cartel, 1926-30, consisted of national steel cartels united in an association. The national steel cartels had government support, but were primarily of a private character.

This first cartel produced 30 percent of the world's output of steel and 66 percent of world exports. It collapsed, however, in 1930 owing to internal problems. In 1930 a second international steel cartel experienced half a year of frustration. In 1931 a third cartel lasted for only two months. A fourth cartel that lasted from 1933 to 1939 had, according to Stocking and Watkins, some success in keeping prices higher than otherwise would have been the case and was also able to discriminate between customers. The price series does not, however, seem to support a judgment on the cartel as being efficient.

Tea

There have been a number of attempts to organize cartels in the tea industry. The International Tea Cartel from 1933 to 1939 was regarded as an interesting example of a collective marketing control established by trade associations with the cooperation of governments. The concentration in the industry was low. Demand was probably price-inelastic, as is the case today according to C. Fred Bergsten. Demand also seems to have been income-elastic in the relevant period. Cost differences were most likely high. The war prompted the British Ministry of Food to take over the whole tea supply and fix prices according to the average price prevailing at the end of 1938. The price series seem to indicate that the cartel had no effect on prices.

Sugar (1864-1939)

The concentration in the sugar industry is low. In the export markets, however, the concentration is high due to common sales agencies. According to Stocking and Watkins, demand was price-inelastic prior to World War II. Demand seems to have been income-elastic in the same period.

The 1864, 1902-12, 1929, 1942, 1953, 1956, and 1958 cartel attempts in this industry are included in sample 2. The first international sugar cartel we include in this sample is the so-called Chadbourne Agreement of 1931-35, which was a private marketing control agreement, approved and enforced by the respective governments. Failure to restrict production efficiently and the rapidly increasing market share of outsiders made the Chadbourne Agreement collapse. On the initiative of the League of Nations, a new international agreement was signed on May 6, 1937. It was a diplomatic treaty between twenty-one governments representing 85-90 percent of the world's sugar production and consumption. Prices were stabilized some 30 percent above the 1935-36 average prices, and the cartel was accordingly judged inefficient. The agreement was disrupted by the war in 1939.

Sugar (1958-61)

Today nearly 90 percent of the world's sugar is either consumed in the areas where it is produced or is marketed under a quota system. This means that a very small proportion of all sugar produced is freely traded in international markets. In the short-term, corn syrup and other sweeteners can be substituted for cane or beet sugar. The precise elasticity of demand is not well known, but it was judged to be inelastic in the near term.

Sugar trading receives protection from many government-backed commodity agreements. In the U.S. there is a U.S. Sugar Act. In Great Britain the comparable pact is the British Commonwealth Sugar Agreement. In 1958 an International Sugar Agreement (ISA) was negotiated between all of the large producing nations in order to stabilize the wide fluctuations in prices. This international agreement was not able to restrict fluctuations, but it did serve to prevent any further declines in average prices. The ISA broke up in 1961 because of growing difficulties between the U.S. and its major sugar trading partner, Cuba. Until that time the U.S. had gotten 75 percent of its imports from Cuba. However, in mid-1961 the U.S. cancelled all international trade with Cuba and sought other sources of sugar elsewhere in Latin America. At the same time Cuba had huge supplies which had to be sold in other, non-U.S. markets. This instability in market conditions was enough to cause the ISA to crumble and world prices to fall.

Copper (1950-70)

Most of the free world's copper supply is found in fewer than seven countries and is refined by what is known as the "big eight" firms, for uses in electrical and other industrial processes. Quantitative estimates of the short-run elasticity of demand (between .21 and .48) have underscored that demand is relatively inelastic since not many short-run substitutes are available. In the long run (ten years or more), alternatives are more feasible and demand is relatively elastic (approximately 2.8).

During the mid-1950s, and again in the mid-1960s, producers made attempts to influence the market price. These actions were generally taken with the full knowledge and cooperation of the respective governments. Chile, Peru, Zambia, and the Congo have been the most active in this regard and have formed a joint body, CIPEC, to promote their common interests. The initial price experiment (1955-56) was undertaken by a Zambian producer who felt that he could appreciably affect the price of copper by imposing a ceiling on price. The unilateral attempt was unsuccessful, however, for the cooperation of other producing firms was not attained.

A second price experiment (1965-66) found more support among the large producers, and consequently was far more successful from their perspective. In the two-year period, copper prices doubled as the "big eight," as well as smaller firms, temporarily agreed on common goals. After two years of steadily rising prices, agreement among producers faded as some began shading on prices. Explanations of the breakdown have noted that some of the less developed countries have vastly different time horizons than many of the private producers. For example, while Chile was interested in exploiting a short-run demand inelasticity, many of the private firms were much more conservatively inclined, with an eye toward preserving long-run demand and discouraging the development of copper substitutes.

SAMPLE 2

Sample 2 consists of the industries on which we were able to obtain information sufficiently detailed only to code the five-dimensional cartel table defined above. The sources of information are identical to those of sample 1. The influence of our personal judgment is, however, more severe on this sample than on the first sample. The results are shown in Table 3-3.

Wheat

In 1933 the first international wheat agreement was established by governments of wheat producing and importing countries, without

Table 3-3.   Characteristics of Sample 2 Cartels

Year	Copper 1918	Copper 1929	Quebracho 1819	Quebracho 1926	Quebracho 1934	Sulfur 1934	Potash 1926	Phosphate Rock 1933	Magnesium 1923	Iodine 1878	Diamonds 1930
1. Length of Formal Agreement	6	3	3	5	8	6	13	6	18	61	12
2. Concentration of Production	2	2	2	2	2	2	2	0	0	0	2
3. Cartel Members' Share of Total Production	2	2	2	2	2	2	2	2	2	2	2
4. Breakdown Externally (?)	0	1	0	0	0	1	1	1	1	1	1
5. Efficiency	0	1	1	1	1	1	1	1	1	1	1

Year	Sugar 1864	Sugar 1902	Sugar 1929	Sugar 1942	Sugar 1953	Wheat 1933	Wheat 1942	Wheat 1949	Wheat 1959	Platinum 1918	Platinum 1931	Salt Lake 1926	Salt Lake 1930	Coffee 1957	Coffee 1958	Coffee 1959
1.	0	10	0	0	0	1	5	10	3	0	2	4	9	1	1	3
2.	0	0	0	0	0	0	0	0	0	2	2	0	0	2	2	2
3.	0	0	0	2	2	2	1	1	2	2	2	0	0	1	1	2
4.	0	0	0	0	0	0	0	0	0	0	1	1	1	1	1	0
5.	0	0	0	0	0	0	0	0	0	0	0	0	0	0	0	0

direct reference to private entrepreneurs or their organizations. The agreement broke down within a year owing to disagreement over quotas and acreage reduction in addition to a very unfavorable price development. In 1942 Argentina, Australia, England, the U.S., and Canada established a new pool, limited in scope, but to be extended after the war. This plan collapsed, however, in 1947 when Argentina abstained.

The postwar international wheat arrangements have been for three-year periods. The 1949 wheat agreement was renewed in 1953 and 1956, then revised substantially in 1959 and renewed in 1962, the last year on which we have any information. Too weak jurisdiction over members has made these agreements inefficient.

Copper (1918-40)

In 1918 a cartel was formed to liquidate the tremendous stocks of copper piled up as a result of the war and to regulate new production and exports. It was wholly American in membership. It represented 95 percent of the American production. The only outsider was Katanga, still in its infancy. The cartel was disbanded in 1924 after dissension arose between the companies with foreign properties and those with purely domestic properties. The cartel was successful in liquidating stocks without causing a sharp fall in prices, and also in regulating exports. It was consequently judged to have been efficient.

In 1926 Copper Exporters Inc. (a Webb-Pomerene association) was established. The company controlled 95 percent of the world's production of copper. The combined effect of the 1928-29 boom and cartel rationing sent prices upwards. The resentment against the cartel grew so strong, however, that a buyer's strike was called. From then until the dissolution of the cartel in 1932, with the enactment of the U.S. excise tax on copper, the power position of the cartel steadily declined. Regarding the 1935-41 international copper cartel, information relating to world markets outside the U.S. is scarce.

Platinum

In 1918 several producers tried unsuccessfully to organize a cartel. In 1931, however, an agreement was signed, only to break down in 1933. Because platinum is mainly a by-product and palladium, a substitute, was not included, control of the market by the cartel seems to have been impossible.

Quebracho

Argentina and Paraguay have completely dominated this industry. In both countries the quebracho producers were organized in a government-sponsored cartel. In the periods 1919-22, 1926-31, and 1934-46 (1946 being the last year on which we have information), these two national cartels operated jointly in the international market by establishing exclusive sales agencies, export quotas, and uniform price policies. In 1942 the American agents were indicted for violation of antitrust regulations. As we have not been able to obtain additional information, this indictment (as well as a 1920-39 price series) is the basis on which we have judged the cartels to have been efficient.

Sulfur

In 1838 the United Kingdom broke the Sicilian sulfur monopoly by sending gunboats. In 1934 a cartel was organized among the U.S. and Italian producers. The U.S. had at that time 80 percent, Italy 11 percent, and Japan 6 percent of the world's production of crude sulfur. The cartel had complete control over export supplies and markets through the use of export quotas and uniform prices. According to Hexner: "Significant international agreements concerning sulfur are most characteristic of modern cartellization." U.S. antitrust actions and some information on prices is the basis for judging the cartel to have been efficient up until World War II.

Sodium Sulphate (Salt Lake)

Important outsiders seem to have made life difficult for the cartels in this industry from 1926-30 and 1930-39.

Potash

Under strong pressure from the French and German governments, the potash exporters of these two countries formed a cartel in 1926. Germany was at that time responsible for about 60 percent and France for about 16 percent of the world's production of potash. Export prices were to be determined by production costs. American producers were, however, indicted in 1939 under the Sherman Act because of alleged cooperation in price policies among themselves and with the European cartel. It was stated that this natural monopoly was abused by Germany and France. As export prices of potash were not published, the above-mentioned evidence is the basis for judging the cartel to have been efficient.

Phosphate Rock

World phosphate exports were regulated by an agreement established in 1933 and further amplified in 1934 and 1935. The agreement embraced the whole international market. The agreement is surrounded by a high degree of secrecy. From 1929 to 1939 phosphate prices tend, however, to support our judgment on the cartel as having been efficient.

Magnesite

In 1923, Czechoslovakian and Austrian producers established a joint-stock sales company to regulate the international magnesite market. An "understanding" with American producers was also obtained. The large magnesite consumers were the shareholders of the magnesite companies involved. In 1941 there was a U.S. Justice Department indictment for U.S.-European division of world magnesite markets. On this basis we judged the cartel to have been efficient.

Diamonds

Government licensing and monopoly support have helped monopolize the diamond industry. In 1930 a diamond-trading company was established as the sole selling agency for 99 percent of African diamond production, or 95 percent of world diamond production. The British government took over the company in 1942, after what is assumed to have been twelve successful years.

Coffee (1957, 1958, 1959-62)

Coffee is primarily grown in Brazil, other Latin American nations, and Africa. Since World War II, world production has sharply increased, while Brazil's market share has steadily declined. Production is almost universally undertaken in the less developed countries and as such represents a substantial amount of these countries' GNP. Because of chronic oversupply, especially in Brazil, several exporting nations have periodically attempted to stabilize or bolster sagging coffee prices.

In 1957, and again in 1958, Latin American coffee agreements were signed. Most Latin producers agreed to hold back a percentage of their harvests from the market, with Brazil leading the endeavor with a 40 percent reduction. Neither agreement was successful in raising prices because African nations filled the gap with their own coffee.

In 1959 the African producers agreed to enter an international coffee agreement, with 85 percent participation by world producers. The agreement set fixed export quotas based on 90 percent of past exports or 88 percent of estimated future exports. The agreement was renewed annually and was significant in that consuming nations were also included. The system has had the effect of providing a floor and increased stability for formerly volatile coffee prices.

CONCLUSIONS AND EXTENSIONS

There are nine efficient and fourteen inefficient cartels in sample 1. Also, ten efficient and eighteen inefficient agreements make up sample 2. Therefore, of the fifty-one significant cartel organizations, only nineteen achieved price controls that raised the level of charges to consumers significantly above what they would have been in the absence of the agreements (Tables 3-4 and 3-5).

The efficient cartels did not seem to last very long. Although formal organizational agreements (to set up cartel management, for example) lasted longer in the efficient cartels, the average length of effective controls on price was not more than four to five years. The

Table 3-4. Summary Table—Sample 1

		Efficient	*Inefficient*
A.	Number of Cartels	9	14
B.	Average Length of Formal Agreement (Years)	5	3.1
1.	Concentration of Production (High:1, Low:0)	0.9	0.36
2.	Concentration of Exports-Imports (High:1, Low:0)	0.9	0.5
3.	Demand Elasticity (Elastic:1, Inelastic:0)	0.22	0.06
4.	Income Elasticity (Elastic:1, Inelastic:0)	1	0.78
5.	Short-Term Substitutes (No:0, Yes:1)	0.22	0.43
6.	Long-Term Substitutes (No:0, Yes:1)	0.77	0.43
7.	Government Involvement (No:0, Yes:1)	0.11	0.58
9.	Cartel Members' Share of Total Production (Very high: 2, High:1, Low:0)	0.9	1.14
10.	Cartel Members' Share of Exp./Imp. (Very high:2, High:1, Low:0)	2	1.58
12.	Cost Differences Among Cartel Members (High:0, Low:1)	0.9	0.58
14.	Potential Time Horizons of Agreements (Long:0, Short:1)	0	0.08
15.	Breakdown (Nonmarket related:0, Market-related:1)	0.66	0.70
16.	If Market-related Breakdown, Then (Externally:1, Internally:0)	0.33	0.30
17.	If External Reason for Breakdown, Then (Supply:1, Demand:0)	1	0.5

Table 3-5.   Summary Table—Sample 2

		Efficient	*Inefficient*
A.	Number of Cartels	10	18
B.	Average Length of Formal Agreement (Years)	8	2.7
2.	Concentration of Production (Very High:2, High:1, Low:0)	1.6	0.55
3.	Cartel Members' Share of Total Production (Very High:2, High:1, Low:0)	2	1.2
4.	Breakdown (Externally:1, Internally:0)	0.6	0.28

mercury cartel in the 1930s and 1940s and the potash, magnesite, and diamond cartels of the 1930s seem to have been able to control prices for as long as a decade, but these were not major products in international trade. The more important products, such as rubber in the 1930s or aluminum, copper or sulfur before World War II, experienced cartel longevity from one to four years.

A number of factors are important in the longevity of the efficient cartel. Without these factors, it would seem to have been impossible for most cartel organizations to last for more than a few months.

1. *Concentration of production* was characteristic of the efficient cartel. Approximately 90 percent of the efficient cartels in sample 1 had concentration levels higher than 50 percent (the largest four firms had more than 50 percent of total production or capacity to produce); but only 36 percent of the inefficient cartels had concentration levels this high. Similarly, the efficient cartels controlled a very high percentage of exports.

2. *Demand conditions* also strongly affected the chances that the cartel agreement worked well and lasted for a reasonable period of time. The summary tables show that the efficient cartels were characterized by inelastic demands (lack of sensitivity of quantities demanded to price changes), and that they also were characterized by the lack of short-term substitutes in most cases (only 22 percent of the efficient cartels in the first sample had no long-term substitutes); but this was also true of the inefficient cartels. The ability to substitute other products in the long run may have limited both the length and efficiency of the agreement.

3. *Government involvement* made a difference in the success of the agreement. Government agencies were involved in the organization of the cartel in almost 60 percent of the cases in which the cartel did not work well. Although not much information was

provided in the studies as to what the governments' activities were, it is presumed that at some stage political and diplomatic relations entered into the cartel organizations so as to break down the agreements.

4. *Supply* conditions differentiated efficient from inefficient cartels. Most of the successful cartels had as members one or two firms with production costs much lower than other firms, the lowest-cost firms tending to "dominate" operation of the agreements. When cartels did break down, it was mostly because of entry of additional suppliers or the expansion of supply by small firms outside the cartel's agreements (as shown by line 14 of Table 3-4).

In summary, there seem to be several important factors differentiating efficient from inefficient cartel agreements.

Cartels were able to raise prices for four years or more, where *concentration* or production was high, *demands inelastic*, and where *few short-term substitutes* were available for the cartelized product. Governments were involved in breaking down agreements. Operating *cost advantages* and the presence of few outside sources of supply able to expand capacity were important for cartel success. These factors are shown in the summary table for sample 1, as those conditions of the fourteen listed, for which the efficient cartel had significantly different values from the inefficient cartel.[a]

Much the same is shown by sample 2, because the *concentration* of efficient cartels is significantly higher than the inefficient. Also, the cartel members' *share of total production* was much higher, and if cartel breakdown occurred, it was mostly because of entry into international markets by new firms.

There are further important dimensions not included in the findings from the earlier research studies. Indications scattered throughout the studies are that an important additional factor for cartel success or failure is tight control of distribution channels. The iodine cartel lasted more than fifty years as an organization, without significant disruption, by making all iodine sales out of a single cartel association office in London (although there were no findings on the ability of this organization to raise unit price above unit cost). There are other examples in which additional elements of control seem to have followed from cartel supervision of distribution, but these are too scattered to lead to a research conclusion at this time. Similarly,

---

[a]By "significant difference" we mean a rough qualitative difference in the magnitude of the statistics between 0.0 and 1.0 in the two columns of the tables. For those six factors termed "significant," the differences in table values range from .32 to .66. Although there are smaller differences indicated by other factors, we chose to ignore them at this time because of small sample size and the highly qualitative nature of the values assigned between 0 and 1 between each cartel attribute.

the level of *concentration among consumers* seems to be important in some cases. Where there are only a few consumers and they are able to play one cartel member off against the other, then the efficiency of the cartel would appear to have been limited. But high buyer concentration was found only in very few cases and cannot be said to be a "finding" from the research analysis.

Probably the most important determinant of the longevity of a cartel agreement is the way production and profits are allocated among the cartel members. The unavailability of information on this aspect of a cartel agreement made it impossible to determine the level of conflict among the cartel members. Given that the "efficient" cartel broke down more often because of the emergence of competition among the members rather than because of the response of nonmembers, the internal operating mechanisms of cartels have to be analyzed if we want to learn more about the stability of cartel-dominated markets.

The conclusions on important factors for cartel's success, and the summary tables themselves, are based upon the reading and evaluation of research materials in a wide variety of industries and cases. There is a strong element of personal judgment in the assigning of such attributes as "high concentration" or "lack of short-term substitutes." It should be stressed that another review of this material might well establish somewhat different factors in the efficiency of agreements, or whether in fact an agreement was efficient or inefficient. But the over-all impression that *efficient cartels do not last very long* would probably not be dispelled. Neither would the finding that high concentration, the presence of a dominant producer, and the lack of expansion by those outside the cartel contribute very strongly to cartel price control over the four- to six-year lifetime of a typical organization. The review of existing material on international commodity cartels also leaves the impression that in industries whose structure is favorable for collusive arrangements successful cartels tend to come back.

IMPLICATIONS FOR OPEC

What do these factors tell us about the causes for the efficiency and longevity of the present-day petroleum cartel? There have been petroleum cartels at an earlier time; the "as is," or "Achnacarry," agreement of the late 1920s to maintain output shares of American oil-exporting companies collapsed in 1930 without having had a significant effect on European markets. Later, similar agreements with quotas and fines did not collapse, but there is little or no evidence that they had an appreciable effect on price levels before World War II. From 1945 to 1960 there were no formal agreements.

But prices were "high" in the sense that marginal production costs plus user charges could not have exceeded $1.00 per barrel, while prices were mostly centered around $2.00/barrel. The companies were thus able to maintain a noncompetitive price level. Since the advent of the highly efficient OPEC cartel operation in the early 1970s, price-cost differences have increased to many times those expected from the earlier cartels.

The present-day OPEC agreement has many of the characteristics found in the earlier cartels that were successful for limited time periods in other industries. The demand for its product is inelastic, and there are few short-term substitutes for this product. Concentration within the cartel is substantial, and OPEC itself as an organization supplies about 90 percent of the total flow in international trade. The Arab subset of OPEC supplies 54 percent alone of the total international flow. There are substantial cost differences among firms, with the Persian Gulf producers having significant cost advantages and significantly greater capacity than the "fringe" of Southeast Asian, East African, and South American countries.

The Teheran and Tripoli agreements in 1971 between the oil-exporting countries and the oil companies may be considered as the first evidence of an efficient producer-country petroleum cartel. Since then the producer countries have been able to raise prices successfully to a level that makes OPEC the most efficient cartel in modern times. The OPEC countries have not, however, been able to agree on and stick to a formal system for sharing production among the member nations. As long as the OPEC members accept the way the major oil companies allocate the reductions in production due to the higher prices and the world recession, the intracartel level of conflict can be kept at a minimum. The lack of formal production and/or profits allocation systems makes, however, OPEC as vulnerable to emergence of internal competition as the cartels that have been reviewed, even if the willingness to accept production cutbacks and to live with a huge excess capacity has been impressive.

If OPEC were to follow the pattern set by the nineteen earlier "efficient" organizations, then it would likely have a four-to-six-year duration. The primary source of breakdown of price controls would likely be the significant additions of supply from either the "fringe" of OPEC members, or the nonmember countries (in this case, the North Sea countries, Canada, and the United States) which by self-supply reduce the demands placed on the low-cost Persian Gulf states. We would, however, expect to observe some successful post-OPEC cartel-like organizations dominate the international petroleum market for four-to-six-year periods.

Chapter Four

Cartel Theory and Its
Relevance to the
International
Petroleum Market

OLIGOPOLY THEORY

A cartel is a group of suppliers that has entered into an explicit agreement to limit competition for mutual benefit.

This is feasible when the number of suppliers of a particular good or service is "small," and each supplier is aware that his profits depend on the behavior of each other supplier in the industry. That is, when the market structure is oligopolistic. A cartel is an attempt to find a cooperative equilibrium in an oligopolistic market, and is thus a special case in oligopoly theory.

Oligopoly theory is intended to help us understand the behavior of markets that are structurally located somewhere in between perfectly competitive markets and markets dominated by a single monopoly. The distance between these two extremes is huge, and there is consequently room for a number of theories to explain the infinite number of points between the two polar cases. The lack of success of the simple theories on the relationship between market structure and behavior, theories designed to explain the whole continuum of possible market structures and behaviors, has given rise to a bewildering number of theories based on a priori behavioral assumptions built into formal models or on case study information. The closer the market structure assumed or observed is to one of the two polar cases, the stronger and more plausible seem the conclusions. Conclusions with respect to the dark area in between the polar cases are more a commonsense extension of the implication of the polar cases, rather than the outcome of a generally accepted, adequate theory of oligopoly. Excellent reviews of the state of the art of oligopoly theory are provided elsewhere [10,17].

Oligopolistic markets have provided a rich source of hypothesis to be explored in formal models by the mathematically inclined, or to be evaluated in case studies by patient empirically-oriented researchers. The classic models of Cournot, Bertrand, Edgeworth, and von Stackelberg, in which the reaction of each oligopolist to the action of each other oligopolist is specified, constitute the core of the formal models exploring the behavioral interdependencies between profit-maximizing firms. Bishop's more recent work on the nature of oligopolistic warfare also explores the behavioral interdependencies between oligopolists [4]. By disregarding the behavioral complexities within the existing oligopolistic market, the "limit-price" modelers simplify the world into "existing firms" and "potential entrants." Such a simplification makes it possible to focus on the interaction between the ways in which potential entrants perceive the response to entry by existing firms and the behavior of the existing firms. Harrod, Edwards, Bain, Sylos-Labini, Modigliani, and Bhagwati have among others developed this formal modelling approach. Work in this tradition by Gaskins, Kamien and Schwartz, and Baron also includes an explicit treatment of uncertainty and of the dynamics of the entry process. The above-mentioned "limit-price-model" simplification implies that this modelling approach should be regarded as belonging to the family of monopoly models, rather than being included in the portfolio of oligopolistic models.

The formal models have not, however, proven very useful to students of particular markets, and their existence seems to have had a negligible effect on applied research as exemplified by the review of research on international commodity cartels. The lack of applicability of formal models to the analysis of particular markets has given rise to more story-telling on oligopolistic markets than on any other subject in economic theory. Detailed studies of individual industries and firms, studies on how actual firms make actual decisions, have provided the basis for our insight into oligopolistic markets more than the formal models. The important characteristics of oligopoly behavior are not captured by the conventional models.

More recent efforts have been made to construct simulation models that incorporate elements of story-telling and of theoretical studies in one analytic framework [13,16]. Such models may possess some of the rigor of the formal modelling tradition without giving up the richness of the story-telling approach.

CARTEL BEHAVIOR

The existence of OPEC as a formal organization to coordinate the oil policies of the member states, to stabilize the price of oil, and to assure a stable income to the producer countries makes it logical to focus on the kinds of explicit collusive agreements that are labeled cartel agreements.

The economic incentive to organize and operate a cartel is that each individual member of an industry can make larger profits by receiving an appropriate share of the industrywide monopoly profits than by following any other market strategy. The benefits of being a member, rather than the disadvantages of being a nonmember, are the incentive to participate in a cartel.

Legal and political constraints may, however, make it impossible for a cartel to reach the monopoly solution, in which case a cartel will not necessarily be the profit-maximizing strategy for all suppliers in an industry.

The existence of a cartel agreement does not solve the problem of lack of a unique solution concept inherent in oligopolistic markets. The lack of a unique profit-maximizing solution tends to make cartel agreements unstable. Contributing to the observed instability of cartels is the fact that in most countries collusive agreements constitute criminal conspiracy. In the international domain the sovereignty of the nation-state makes it impossible to legally enforce or prosecute collusive agreements initiated by intergovernment action. The cartel agreements of interest are consequently those that cannot be controlled by court-enforced sanctions. It is no trivial task to discourage individual cartel members from price-cheating to capture additional sales and profits when threatening believable and punitive retaliation is the only means of enforcement.

The ease with which information is obtained about what rival firms are doing and what potential entrants might do is dependent on the number of firms in an industry, and the similarity of the firms, including similarity with respect to the perceptions of the present and to expectations about the future. The smaller are the costs associated with gathering information on the activities of rival firms, the smaller are the costs associated with enforcing a collusive agreement, and the greater will be the ability of a group of firms to behave as a single monopoly.

If the colluding group of firms is heterogenous, and the demand for the industry's output is inelastic, whereas the demand for any individual firm's output is highly elastic, and the marginal production costs are small compared to the cartel price, then there are considerable incentives for each individual member to engage in price-cheating. The same set of characteristics also provides, however, strong incentives for a group of firms to form a collusive agreement. The logical implication of the different effects of the above-mentioned incentives on individual and on joint behavior should be a fluctuating market price. If the cartel is reorganized every time cheating is detected, then the individual firms would be tempted to make "a fast buck" before being discovered, and then return to the old routine. Such behavior would imply ever increasing fluctuations over time.

CARTEL MANAGEMENT

The tasks of the managers of a cartel agreement are: (1) to design a pricing strategy that will maximize the profits of the membership, (2) to allocate production in the most efficient way, (3) to design a system for allocating joint profits in a "fair" way, and (4) to police the cartel agreement so that no one can chisel to his own benefit and to the detriment of some other cartel member.

Here we will primarily be concerned with the first three tasks. The fourth will be briefly considered to indicate the nature of this problem. The policing problem is a source of instability, and may as such be used as an argument in a "story" about cartel breakdown. The policing problem is, however, considered beyond the scope of the market characteristics included in the model described in Chapter Five. The pricing problem and the possible systems for allocating production and profits are explicitly included in the model.

It has been claimed that the profit-maximizing price for a cartel is different from the price a monopolist with the same number of plants would have charged [14]. My model assumes that the "optimal" cartel price is the "monopoly" price. A cartel should never fiddle with the price once the joint-profit-maximizing price path has been reached.

It is my hypothesis that the reason for cartel breakdowns is the lack of appropriate systems for allocating production and profits. The center of attention when studying the stability of a cartel-dominated market should therefore be the division of profits. There are more profits to be shared by the members of an industry when the joint-profit-maximizing solution is chosen than in any other case. A discussion of possible ways to divide profits so that nobody will feel tempted to break the cartel policy for economic reasons is therefore appropriate. In its most general form the problem can be stated as for the "industry-to-be-cartelized" mentioned below.

The demand for the industry's output is

$$Q = Q(P) \qquad (4.1)$$

The number of suppliers in this industry is fixed by an absolute barrier to entry. The total cost, C, of producing the industry output depends on the level of output, Q, as well as the distribution of production, W, among the firms:

$$C = C(W,Q) \qquad (4.2)$$

The joint-profit-maximizing solution is that of a multiplant monopolist, i.e., the price, quantity, and distribution of production resulting from equalizing marginal revenue with marginal cost in the plants being operated to produce the monopoly quantity. If an industry behaves like a multiplant monopolist, the level of profits generated will be large enough to compensate all the members sufficiently to make price-cheating unattractive. If we stick to the rule that the loyal members of a cartel will choose the price and quantity that will maximize their joint profits, it is possible to design a system for division of profits that will make the potential cheater's profits at least as large by sticking to the cartel policy as by behaving as a price-taker.

If we let π_i^* denote the profits of a cheater when behaving as a price-taker with the loyal firms reacting to the defection by maximizing profits, and π^* denotes the joint industry profits in the case of perfect collusion, the optimal system of division of profits, WP, is such that:

$$WP_i \ \pi^* \geqslant \pi_i^* \qquad (4.3a)$$

or

$$WP_i \geqslant \frac{\pi_i^*}{\pi^*} \qquad (4.3b)$$

This system of division of profits removes the economic incentive to chisel. It can easily be shown that the loyal firms of a cartel are always better off by compensating a potential cheater in this way than by letting him defect and then maximizing profits without him. If both the potential cheater and the loyal firms are better off under this profit-sharing arrangement, everybody is better off.

The economic incentive to breaking the collusive contract is therefore removed.[1]

The general conclusion with respect to the use of price and quotas for stabilizing purposes is that a cartel should focus on how to allocate monopoly profits to remove incentives to cheating and never fiddle with the joint-profit-maximizing price. Conversely, inability to allocate the monopoly profit and bribe potential violators forces the cartel to operate on the price, with possible destabilizing effects.

CONCLUSION

This book focuses on a particular market, the international petroleum market. The existing institutional environment of that market makes it plausible to focus on cooperative market solutions in general, and on possible cartel coalitions in particular. Given the state of the art of oligopoly models assuming cooperative market behavior, a simple monopoly model is the most plausible when describing the price behavior of a given colluding group. Conventional oligopoly models are not useful when dealing with the problem of which firms are likely to form a collusive agreement under which circumstances. A more detailed study, "story-telling," is needed to hypothesize which coalitions are likely to emerge under the different circumstances the industry may have to face. A specific analysis of the membership of OPEC is needed to construct a plausible set of intracartel reaction functions. The way production and profits is

---

1. If we denote the profits of the loyal firms in the presence of a cheater as π^*_{-i}, it is sufficient to demonstrate that the following inequality holds:

$$(1 - \frac{\pi^*_i}{\pi^*}) \pi^* \geqslant \pi^*_{-i} \tag{4.4a}$$

or

$$\pi^* - \pi^*_i \geqslant \pi^*_{-i} \tag{4.4b}$$

or

$$\pi^* \geqslant \pi^*_{-i} + \pi^*_i. \tag{4.4c}$$

The inequality (4.4c) holds by definition of π^*, which was defined as the maximum profits that could be extracted from the industry. π^* is therefore at least as large as the total profits resulting from any other price-production combination.

allocated among the cartel membership is a good proxy for a measure of the desirability, and hence also of the feasibility, of various cartel compositions.

The need to combine formal modelling aspects with informal story-telling makes a simulation model the most ambitious approach that can be made to the analysis of a particular market without sacrificing the empirical validity of the analysis.

Chapter Five

The Model Representation of the International Petroleum Market

The shortcomings of formal models designed to analyze oligopolistic markets imply that the structure of a model designed to analyze a particular oligopolistic market should be flexible enough to combine features of formal modelling with informal "story-telling." As it is our intent to study the behavior of the market as it actually exists rather than calculating "efficient" or "optimal" patterns of market development, and because there are many important time-dependent relations to include, it was decided to develop a simulation model. A simulation model easily permits formulations of a number of exporter decision rules or exporter market strategies. The form of the relationships or of the equations of the model was chosen so that an explicit analytic expression for the market clearing price consistent with each combination of exporter strategies could be derived. No numerical or "optimizing" subroutines are therefore needed to solve for a market-clearing price path. The structure of the model and the functional relationships are constructed to explore the implications of pricing behavior reflecting a monopolistic market structure, a competitive market strategy, or any combinations of the two. The change in the set of decision rules applied in each period makes the model evolutionary in the sense that the price behavior of the market may change from period to period.

There is a complex transportation network serving the international petroleum market, and a number of different crude oil qualities and petroleum products are flowing in international trade. It was decided, however, that the costs in terms of increased model complexity would outweigh the insights gained by explicitly representing this transportation and products network. Oil is treated as a homogenous product and the geographic location of exporters and importers is disregarded. The international petroleum market is thus treated as a "bathtub," therefore the label the "bathtub" model.

THE IMPORTER REGIONS

The international petroleum market is presently the marginal source of energy in the world. The international price of petroleum determines the price that can be charged for all fuels, and thus the quantity supplied of all other fuels. The demand facing the world's exporters is thus a residual demand. The bathtub model consists of such a residual demand framework.

In the model development we deal with a group of net importers, denoted by the index i, and with a set of exporters, j. We define E_i^t as total demand for energy in region i in year t, and let P be a vector of past and present energy prices. The relationship between the international price of petroleum and the total demand for energy in region i in year t can then be represented as

$$E_i^t = E_i\,(P,t). \tag{5.1}$$

The market share of oil, in region i in period t, M_i^t, is likewise a function of past and present prices in the international market,

$$M_i^t = M_i(P,t), \tag{5.2}$$

as is also the indigenous supply of oil in region i in year t, S_i^t, where

$$S_i^t = S_i(P,t). \tag{5.3}$$

The resulting demand for imports to region i in year t, D_i^t, is consequently

$$D_i^t = D_i\,(P,t) = E_i^t \cdot M_i^t - S_i^t \tag{5.4}$$

In the present version of the model, the world is divided into four importing regions, $i = 1, \ldots, 4$. They are Western Europe, Japan, United States, and the rest of the world (not including USSR, Eastern Europe, and China, which are included in the model as net exporters only).

The relationships defined in Equations (5.1) to (5.3) are complex. A simplification of these relationships is required to solve for the market-clearing price and quantity even in the case of perfect competition. As we focus on noncompetitive market behavior, including monopoly behavior, in which case the first derivative of the residual demand function together with the first derivative of the monopoly unit's total cost function determine the market price, even more simplified versions of the relationships defined above are required. To test a number of ways of structuring the behavioral characteristics of the various participants in the international petroleum market, approximations were made to the above relationships such that the system of equations representing the international market could be solved analytically rather than having to solve the system of equations by applying some more costly numerical method. A set of linear approximations to the regional oil/energy relationships was therefore constructed.

Price Expectations

An important characteristic of the oil/energy sector is the long lead times due to the capital intensity of the sector. That is, a number of periods are required to adjust the capital equipment of the oil/energy producing and consuming sectors to a significant change in price. The oil/energy producers and consumers will hence adjust their capital equipment to the market price they expect to prevail when the equipment will be on stream some time in the future. The sophistication of these investors may differ a great deal, and there is consequently room for a number of rules representing how expectations with respect to the future price level are formed. Given the state of the art of predicting the future and a strong desire to simplify as much as possible such expectations rules, the present price was chosen as a first approximation to the expected price. A more complex expression for the expected price is introduced on page 71.

If we define "the long term" to be the number of periods required for a complete adjustment of the capital equipment of the oil/energy sector to a significant change in the price of oil/energy, then the postulated linear approximations and the assumption that the expected price is equal to the present price imply that the long-term relationships between the variables characterizing the oil/energy sector and the past, current, and expected oil/energy price will be of a simple linear form with current price representing the price vector. The long-term total demand for energy in region i in year t, $\overline{E}_i^t$, (where the "bar" above a variable indicates its long-term value), is consequently:

$$\overline{E}^t_i = e1^t_i - e2_i p^t \tag{5.5}$$

Equation (5.5) thus represents the quantity of energy that would have been consumed in a given year and at a given price if energy-consuming equipment could have been adjusted instantaneously to any price level.

The price-slope coefficient, $e2_i$, was simply estimated by calculating the post-adjustment energy consumption level for two different price levels under identical assumptions about income and other effects, and then dividing the absolute difference between the consumption levels by the absolute difference between the two price levels, thereby constructing a linear energy demand function. The time-dependent intercept coefficient, $e1^t_i$, was estimated by calculating an initial intercept in a year in which the actual and the long-term market solution was assumed to coincide (the actual and the long-term solution was assumed to coincide in 1973 at $3), and then extrapolating this initial intercept at a rate such that the level of long-term energy consumption would have increased at the rate of growth of the economy for a given price level. That is, an income-elasticity of one is assumed when calculating the location of the long-term energy demand function in each period.

To get from the simplified long-term relationships to the short-term relationships, the short-term being one year, a process by which the market would move from one long-term equilibrium to another was assumed. This process was assumed to be completely described by the length and form of a distributed lags function. A number of distributed lags functions are frequently used by economists. The reasoning behind the ten-year moving average applied here is very simple. The cost of oil/energy has traditionally constituted only a fraction of total production costs. Even a significant increase in the cost of oil/energy will consequently not make existing plants and equipment instantaneously obsolete; a gradual process of replacement of depreciated assets with "energy-tuned" equipment is more likely. If we assume that a ten-year straight-line depreciation rule is a plausible approximation to the average industrywide way of depreciating and replacing assets, then a ten-year moving average is a plausible approximation to the process by which the economy will adjust itself to a significant change in the price of oil/energy.

In the nonoil energy sector and the oil sector, a number of short-, intermediate-, and long-term measures will be made in response to a significant increase in the price of oil/energy. Secondary and tertiary recovery methods are applied to increase production from existing oil fields in the short term. Opening up of new coal mines and production from new oil fields following increased development drilling are expected in the intermediate term. Nuclear plants and new oil fields will come on stream in the longer term. We may therefore assume that the response process of the energy-producing sector is similar in kind to the process of adjusting energy-consuming equipment. The apparent similarities of the adjustment processes made another simplifying assumption plausible: the same distributed lags function is a plausible approximation to the adjustment process of all the oil/energy related sectors. With the possible exception of Japan, all the regional units are sufficiently aggregated and the number of energy-related options in each region is sufficiently similar, even if the weights given to the various options may differ, that the same distributed lags function also applies to all the consumer regions. That is, the adjustment process of all sectors in all the regions is assumed to be identical.

Linear Market Relationships

By imposing an adjustment process linking the long term to the short term on the long-term demand for energy function, Equation (5.5), the following linear approximation to the relationship between past and present prices and the demand for energy in region i in year t emerges:

$$E_i^t = e1_i^t - e2_i \sum_{k=t-K}^{t} \lambda^k p^k$$

$$\sum_{k=t-K}^{t} \lambda^k = 1.$$

(5.6)

K is the length of the adjustment process and the λ's are the annual weights of the adjustment process. The intercept of this linear relationship is assumed to grow such that energy consumption for a given price will grow at the rate of growth of the economy in region i, G_i^t.[1]

The same procedure was also used to construct short-term market-share and indigenous-supplies relationships, even if the intercepts of the latter two relationships were estimated differently from that of the total demand for energy. In the case of the market-share

---

[1] The equation with a "growth-adjusting" intercept is thus:

$$E_i^t = e1_i^{t-1} + E_i^{t-1} \quad G_i^t - e2_i \sum_{k=t-K}^{t} \lambda^k p^k$$

relationship, the fraction of oil in total demand for energy was also simply linearized. According to the studies made by OECD and FEA, the market share of oil is relatively insensitive to the price of oil within a fairly wide price range [2,8]. The intercept of the linear market-share relationship consequently dominates this relationship. A linear approximation to the relationship between past and present prices and the market share of oil does not have some of the properties we would expect from a market-share relationship. That is, if the price of oil fell to zero in a linear market-share equation, the market share of oil would not automatically converge towards one, and if the price of oil increased to infinity, then the market share of oil would not converge towards zero. Given the insensitivity of the market share of oil to the price of oil within the price range we consider of most relevance from an empirical point of view ($4 to $15 in 1975 dollars), as indicated by the studies of OECD and FEA, the linear market-share approximation seems to be a relatively costless simplification. The time horizon of this study is 1990. In the longer term we would expect the market share of oil to be more price sensitive, and for a study of the longer-term implications of a change in the price of oil, a linear approximation to the market-share relationship would be inappropriate.

When an initial market-share intercept, $m0_i$, had been calculated according to the procedure outlined above in the text to Equation (5.5), then an exogenous growth rate, GM_i, was calculated by solving for the annual rate of change of the market share of oil in the 1972 to 1985 period consistent with OECD's $3 constant price of oil scenario. GM_i represents the expected trend in the market share of oil at continued low energy prices. This assumed growth rate thus tends to bias upwards the market share of oil for prices above $3. The long-term price effect was estimated as was $e2_i$ in Equation (5.5) and is incorporated in the slope $m2_i$. The linear approximation to the market share of oil in region i in year t is consequently:

$$M_i^t = m0_i (1 + GM_i)^t - m2_i \sum_{k=t-K}^{t} \lambda^k p^k. \qquad (5.7)$$

Indigenous Supplies

When a promising geological basin is discovered or when the price of oil suddenly increases, we would expect the oil industry to go through three stages, the first being an intense exploration and development effort to get production facilities on stream. Once the production facilities are on stream, we would expect a higher level of production to be sustained for some years before the newly developed fields are being gradually depleted. If the higher price level results in a high rate of discoveries, aggregate production does not necessarily have to fall. In the linear approximation to the indigenous supply function, it was assumed that the price would affect the future level of plateau production in the various regions. It was further assumed that development and gradual starting up of new production facilities, the length of the plateau production period, as well as the rate of decline of production following the period of peak production could be represented by extrapolating the intercept at different exogenous rates over the three typical production stages. The initial equilibrium intercept, $s0_i$, is hence being extrapolated at a price-independent rate of change, d_i^t, dependent only on the stage in the exploration-production process at which the region is producing. The linear approximation to the short-term relationship between past and present prices and indigenous supplies in region i in year t is:

$$S_i^t = s0_i (1 - d_i^t)^t + s2_i \sum_{k=t-K}^{t} \lambda^k p^k. \qquad (5.8)$$

By summarizing the regional demand for imports, the residual demand facing the exporter in year t, D^t, emerges as:

$$D^t = d_1^t + \sum_{k=t-K}^{t} [-\lambda^k d_2^k p^k + \lambda^k d_3 (p^k)^2] \qquad (5.9)$$

where

$$d_1^t = \sum_i [(e1_i^{t-1} + E_i^{t-1} G_i^t)(m0_i (1 + GM_i)^t) - s0_i (1 - d_i^t)^t]$$

$$d_2^t = \sum_i [e2_i m0_i (1 + GM_i)^t + (e1_i^{t-1} + E_i^{t-1} G^t) m2_i + s2_i]$$

$$d_3 = \sum_i (e2_i m2_i).$$

The international market clearing price can then be determined from the accounting identity of Equation (5.10) below, stating that the price has to clear the market in the sense that at that price the level of exports in period t, S^t, is equal to the level of imports in the same period:

$$D^t = S^t. \qquad (5.10)$$

As the focus of this study is on the behavioral options open to the world's exporters rather than the policies of the importer governments, Equation (5.10) is given in its "pure" form. That is, no policy instruments are explicitly included in the equation representing the worldwide demand for imports. To incorporate and to assess the traditional importer-country policy options like tariffs, quotas, consumer taxes, and producer subsidies, only minor modifications of the representation of the consumer regions are required.

THE EXPORTERS

The behavioral options open to the world's oil exporters are represented by a set of pricing or production strategies that the exporters may choose to, or have to, follow on an individual or a joint basis. An explicit analytic expression for the market clearing price or quantity consistent with the various exporter strategies is deduced.

Two versions of the bathtub model have been constructed. To assess the significance or the market implications of the behavior of some stable collusive combinations of exporter countries, a simple version of the bathtub model was constructed. In this simple version of the model, one exporter unit only is explicitly represented. The composition of the exporting group may be changed, however, by transferring countries from the exporter unit to the unit representing the rest of the world's indigenous suppliers, or vice versa. Indigenous suppliers are assumed to behave as price-takers. It is thereby possible to combine price-taker behavior on the part of some exporters with any of the below-mentioned strategies to be played by the other exporters even in this simple model version.

The heterogeneity of the exporting countries and the theoretical as well as empirical instability of oligopolistic markets imply that it is unlikely that all the exporters will follow the same strategy. A more complex version of the bathtub model, the cartel version, was therefore constructed to trace the evolution of the international petroleum market resulting from the changing behavior of the major exporter units. By allowing four exporter units to change decision rules in any given period, the cartel version of the "bathtub" model covers a very wide range of possible market solutions. A number of interesting experiments could be performed, however, by using the simple model version.

Static Perfect Competition
"Static perfect competition" means that the exporter unit behaves

as a myopic price-taker. That is, the exporter unit produces the quantity that equalizes marginal cost and the current price. The marginal cost function is the inverse of the competitive supply function. A linear approximation to the relationship between the competitive supply level from exporter j in period t, S_j^t, and past and present prices was constructed in the same way as the linear approximations to the supply functions indigenous to the regional consumer units:

$$S_j^t = s0_j (1 - d_j^t)^t + s2_j \sum_{l=t-L}^{t} (\delta^l p^l) \qquad (5.11)$$

where $\sum_{l=t-L}^{t} \delta^i = 1.$

Each set of assumptions about the behavior of the exporters generates a separate market-clearing price. The expression for the market-clearing price in the case of "static perfect competition" as well as in each of the other cases described below is deducted in Appendix A.

Static Monopoly

The static monopoly strategy is to charge the price that will equalize the marginal revenue of the long-term residual demand function, $\overline{MR}^t$, with long-term marginal production costs, $\overline{MC}^t$. That is, the static monopolist disregards the form of the adjustment process of the consumer markets and also user cost. The monopolist does, however, exploit the inertia of the consumer markets by extrapolating the intercept of the residual demand function at the anticipated rate of growth of the economy K periods hence, where K is the number of periods needed for adjustment of the consumer markets, and then choosing the current price to be the price that equalizes marginal revenue and marginal cost in this enlarged future market. This pricing rule is labeled "static monopoly" because it is similar to the pricing rule a monopolist would follow in a market that adjusts instantaneously to a change in price and where there are no complications resulting from resource limitations.

The static monopoly pricing rule defined here as well as the other decision rules specified in the following are not deduced from maximization on the part of the exporter. The decision rules are plausible approximating rules, given our knowledge of the objectives and of the likely behavior of the exporters. Actual revenue optimization would probably imply more exploitation of the inelasticity of the short-term demand function; that is, a higher price in the period following the monopolization of the market and then a decreasing price converging towards some long-term rising monopoly price path would probably generate higher revenues than the monopoly rules of this study. It is considered more likely that the producer countries, in the case of continued collusion, will follow some pricing rule that will result in a smoothly rising price, rather than follow a pricing rule that will deliberately produce price fluctuations to maximize short-term gains or to discourage potential entrants into the oil/energy market. A high rate of world inflation may, however, help OPEC decrease the price in real terms while at the same time keeping its strength image in nominal terms, as inflation has helped other unions that have increased their prices beyond the most favorable level for its membership.

Constructing a set of plausible decision rules on the basis of some knowledge of the objectives of the market participants rather than deducing the decision rules on the basis of a theory of profit-maximizing behavior is the major distinction between the "behavioral" approach made in this study and the traditional studies of the behavior of market participants. The inexistence of a theory of profit-maximizing behavior in oligopolistic markets, and the number of plausible decision rules that might be observed in such markets, make a behavioral rather than a rigorously formal approach more useful.

The analytic expression for the static monopoly price is discussed in Appendix A.

Income Stabilization

Income stabilization simply means that the exporters will supply a quantity, S_3^t, such that the resulting income is equal to some target revenue level, I^t.

$$S_3 = \frac{I^t}{p^t} \tag{5.12}$$

In this case, production is simply being "divided out." A number of the exporting countries both within OPEC and outside have expressed concern over the sheer magnitude of their actual or potential petroleum revenues. It is therefore of interest to explore the implications of some exporters imposing a revenue constraint on their petroleum activities, or phrased differently, of some exporters behaving as on a backward-bending supply curve.

Production Stabilization

A different policy approach to the problem indicated above is to impose directly some restrictions on production. The exporters may decide to produce some fixed quantity in a given period or decide on a fixed production path over time, $D*^t$. Supply S_4^t is then:

$$S_4^t = D*^t \qquad (5.13)$$

and the complexity of our supply analysis would be reduced considerably. The recent experience of the petroleum exporters seem to indicate, however, that countries as individuals get used to making money very fast and that expectations with respect to future earnings increase equally fast. The much-debated financial constraints of 1973 never turned out to be binding.

Target Pricing

When the group of exporters decide on a common pricing strategy over time, this strategy is labeled "target pricing." The target price path, TP^t, may be any price path on which consensus has been reached. The production strategy that goes along with target pricing is to produce whatever the market will take at the target price without worrying about how actual production is allocated among the exporters.

Neither the desired income path of the income-stabilizing strategy, the desired production path of the production-stabilizing strategy, nor the desired price path of the target-pricing strategy are determined inside the model. They are not a result of any analysis within the model, but are designed to enable us to explore the implications or the sensitivity of the market equilibrating price/quantity to these extreme, yet plausible, behavioral strategies.

Exhaustible-Resource Competition

The exporters may try to anticipate all future prices for their nonrenewable resource by estimating the price they will obtain for the last unit of the resource, and then determine their present production so that the price today is the net present value equivalent of the ultimate price [6,7]. The ultimate price is defined as the price of the "backstop technology," PB, which is the cost of producing a substitute product resting on a very abundant resource base. If we define r as the discount factor, then the net present value equivalent of the backstop technology price is uniquely determined if we can determine the number of periods until exhaustion takes place. That a resource base has been exhausted at a given price means that the marginal cost of extracting an additional unit from the resource base is higher than the given price. The resource base is therefore a function of the backstop technology price.

Determining the number of periods until exhaustion, N, implies solving analytically the following equation with respect to T, the exhaustion date:

$$R = \sum_{t}^{T} D^t \qquad (5.14)$$

where R is the level of recoverable reserves consistent with a backstop technology price of PB, and D^t is the quantity consumed in the period, given the fact that the price charged is the net present value equivalent of the backstop technology price. The simplest and most naive way of determining N would be to divide R by D°, the initial level of consumption. This procedure implicitly assumes a perfectly inelastic demand for the resource, and a zero income elasticity. The elasticity assumptions of this study are different from zero. An attempt was therefore made to construct an expression for N that would reflect nonzero elasticity assumptions (section A.6 of Appendix A). Once the number of periods until exhaustion is determined, the market equilibrating price is simply the net present value equivalent of the "backstop technology" price.

Exhaustible-Resource Monopoly

Under static monopoly, the optimal price and production is determined by equalizing marginal revenue and current marginal production costs. Under exhaustible-resource monopoly, the net present value equivalent of the ultimate price is substituted for current marginal production costs. That is, the monopoly price and production are determined by equalizing marginal revenue and the net present value equivalent of the backstop technology price, which is a plausible approximation to the Hotelling condition in the monopoly case, namely that the difference between marginal revenue and marginal cost shall grow at the rate of interest [6], when marginal production costs are "close" to zero.

In the present versions of the model, the monopoly price is chosen to be the higher price of the static monopoly price and the exhaustible-resource monopoly price. The exhaustible-resource monopoly price would be higher than the static monopoly price if the net present value of the backstop technology price, the present opportunity cost of the resource, is higher than marginal production costs.

The price reflecting the higher opportunity cost is thus chosen as the monopoly price. The upper limit for the monopoly price is the price of the backstop technology.

Cartel Pricing

The world's exporters are not a very homogenous group. A model intended to identify and analyze the implications of various oil exporter strategies should therefore incorporate and deal with this heterogeneity. The cartel version of the bathtub model allows individual exporters and exporter subgroups to follow different strategies. The oligopolistic structure of the international petroleum market necessitates an explicit representation of the various strategies that might be designed to allocate production and income among the major oil exporters.

The lack of a unique solution concept for oligopolistic markets necessitates more assumptions than in either the competitive case or in the monopoly case to solve for an equilibrating price. There is no uniquely rational behavior that can be specified for an individual oligopolist, since the most profitable behavior for one seller depends on the response of the others. That oligopolists fully recognize their mutual dependence is not sufficient for a unique solution. A unique solution, or an equilibrating price, implies that expected and actual outcomes are identical. Expectations play a crucial role in oligopolistic markets. To simulate the behavior of price in an oligopolistic market, it is therefore necessary to specify the expected and actual reaction of all other oligopolists to one oligopolist's behavior. With three or more oligopolists, the possible relevance of coalitions among some subgroups constitutes a major complication. The number of possible coalitions increases rapidly with the number of individual oligopolists. To narrow down the range of possible market solutions, the OPEC cartel is collapsed into three individual units and two aggregate units. The individual units are unit one, U_1, consisting of Iraq, Nigeria, Indonesia, and Gabon; unit two, U_2, consisting of Iran, Algeria, Venezuela, and Ecuador; and unit three, U_3, consisting of Saudi Arabia, Kuwait, UAE, and Libya. The reasoning behind the selection of these units is indicated in Chapter Six. The two aggregate units are unit four, U_4, consisting of U_1, U_2, and U_3; and unit five, U_5, consisting of U_2 and U_3. All other exporters are included in a separate unit, the competitive exporter fringe.

The demand for cartel output in period t, RD^t, is the difference between the worldwide demand for imports in period t, D^t, and what the exporter fringe can or wants to supply in period t, SE^t.

$$RD^t = D^t - SE^t \tag{5.15}$$

The exporter fringe may also follow non-price-taker strategies like income or production stabilization, but it is assumed that the fringe will not participate in any explicitly colluding agreements. The composition of the above-mentioned units may, of course, be altered. For the sake of simplicity in the presentation, and to avoid pretending that the possible market solutions defined below are not dependent on the composition of the units, the composition of the units will not be changed in the following.

All the above-mentioned characteristics of the exporter group as a whole apply to each exporter subunit, and hence also to the exporters that may form a collusive agreement to extract a monopoly rent. There are, however, additional characteristics of each cartel member that may or may not influence the location and the stability of a cartel-determined market solution. The production capacity of unit j in period t, C_j^t, which is a function of past and expected prices, P, development and production costs, MC_j, time, t, and any cartel policy on prorationing of capacity, cc,

$$C_j^t = C_j \ (P, MC_j, \ t, \ cc) \qquad (5.16)$$

is likely to influence the behavior of cartel unit j.

The production allocated to unit j in period t, Q_j^t, which is a function of the total demand for cartel output in period t, RD^t, and the cartel-determined quota system in period t, W^t,

$$Q_j^t = Q_j \ (RD^t, \ W^t) \qquad (5.17)$$

as well as the associated production profits, and the income requirements of unit j, are major determinants of the desirability of the cartel solution or of the strength of the disintegrating forces working on the cartel. The income requirements of unit j, IR_j^t, are simply assumed to be a function of time, t.

$$IR_j^t = IR_j \ (t) \qquad (5.18)$$

Quotas. Production and profits may be allocated in a number of ways. The quota systems described below have been chosen on the basis of what might be acceptable to the OPEC membership and on the basis of what may be deduced from economic theory on the way production and profits should be allocated among a colluding group.

The simplest possible way of allocating production and profits is to decide on an historic base year and then fix the future market shares at the base-year level. If we denote the base year 0, then an historic quota system, WH, implies that cartel member j will be allocated the following market share:

$$WH_j = \frac{Q_j^0}{RD^0} \qquad (5.19)$$

Production profits are retained by the individual members. That is, no side payments are being made.

The fact that some OPEC countries have a high need for income may make plausible a quota system based on the income requirements of the membership. That is, the market share allocated to producer j in period t when based on income requirements, WF_j^t, is proportional to his relative share of total cartel income requirements in period t, IR_c^t.

$$WF_j^t = \frac{IR_j^t}{IR_c^t} \qquad (5.20)$$

If the cartel could allocate production as a multiplant monopolist, then production should be allocated such that the marginal cost of producing each cartel member's last unit, MC_j^t, was equalized, and equal to the marginal revenue of the cartel as a whole, MR_c^t. WE_j^t is determined such that

$$MC_j^t = MR_c^t \qquad j = 1, 2, 3. \qquad (5.21)$$

In this multiplant scenario, profits should be allocated according to the economic power of each unit as defined in Chapter Four and in Equation (5.23) below. If we denote the total profits of the colluding group in period t by $\P_c^t$, and the total profits of unit j as an outside price-taker assuming the remaining cartel members maximize. joint profits by $\P_j^t$, then a system of allocation of total cartel profits consistent with the economic power of each member, WP^t, would imply that unit j would receive the following share of total profits:

$$WP_j^t = \frac{\P_j^t}{\P_c^t} \qquad (5.22)$$

As defined in Equation (5.22), the "economic power" quotas will not necessarily add to one. These quotas do not apply to all cartel members. They are assumed to be administered by the dominant producer(s) as a way of bribing nondominant producers to adhere to the cartel policies. The dominant producer(s) will thus keep the residual monopoly rent left over after compensations have been paid. The economic power quotas may also be used as a proxy for a "fair" allocation of production when side payments are not feasible.

The dominant producer may also choose to invest in the cartel countries in which the gap between the income requirements of the country and the income from oil exports is the widest. As the foreign investment program of the dominating producer is designed to increase cartel stability, a foreign investment quota system, WI, based on the difference between the income-requirements-based quotas and the economic-power-based quotas has been designed.

$$WI_j^t = WF_j^t - WP_j^t \qquad (5.23)$$

Under such an internal transfer of funds system, each producer country will receive investable funds in proportion to its income requirements. Neither the economic power quotas nor the investments quotas apply to all cartel members, and neither will they add to one. Both quota systems are designed to increase the attractiveness of cartel membership to the producers that might behave differently.

Price Strategies. Each cartel unit has a separate "monopoly" price, p_j^t, to propose as the "optimal" cartel price, p_c^t. The various "monopoly" prices are calculated as was the "monopoly" price described above. Each "monopoly" price is consequently a function of the backstop technology price, PB, the residual demand function facing the cartel, RD, the expected production quota, W_j, the expected growth rate of production, MG_j, the level of reserves, R_j, the cost conditions, MC_j, the discount factor relevant to unit j, D_j, and time, t.

$$p_j^t = p_j \, (PB, RD, W_j, MG_j, R_j, MC_j, D_j, t) \qquad (5.24)$$

The cartel price, p_c^t, is a function of the various units' "monopoly" price.

$$p_c^t = p_c \, (p_1^t, \ldots, p_5^t) \qquad (5.25)$$

The number of possible cartel prices increases rapidly with the number of alternative quota systems and the number of cartel units. Even this simplified representation of four possible quota systems and of five cartel units implies that fourteen different cartel prices might theoretically emerge.

Uncertainty in Oligopolistic Markets

The lack of a unique solution concept in oligopolistic markets and the resulting theoretical as well as empirical instability of such markets make it naive to expect the present price to be the price that will prevail in the future. It is consequently more realistic to assume that the existing and potential participants in an oligopolistic market will adjust their capital-intensive production and/or consumption equipment to some weighted average of the prices that would emerge under the most likely market strategies to be observed. We may define this weighted average of the most likely future prices to be the expected value of the distribution of future prices, $E(P)$. If we further define the most likely future prices to be P_n, $n = 1, \ldots, N$, and still let the actual price be p, then the expected value of the future price distribution in period t can be defined as in the following equation:

$$E^t(P) = \sum_{n=1}^{N} \alpha_n P_n^t + \beta p^t$$

$$\sum_{n=1}^{N} \alpha_n + \beta = 1.$$

(5.26)

Depending on the weights, α_n and β, $E^t(P)$ can be higher, lower, or equal to the present price.

It was assumed above that the monopoly strategy was to equalize the marginal revenue of the long-term residual demand function with long-term marginal production costs. The price term in the long-term residual demand function was the expected price. That is, the monopolist was charging the "optimal" expected price. The expression for the optimal expected price in this "uncertainty" case is deduced in section A.8 of Appendix A.

Because of the problems involved in estimating the coefficients of Equation (5.26), the αs and the β, we will be restricted to a set of judgmental values for these coefficients. The price/quantity effect of uncertainty may, however, be demonstrated through the simple procedure outlined above.

This pricing strategy implies that the monopolist exploits the uncertainty inherent in the oligopolistic market structure. The fact that noncartel market participants expect the cartel to fall apart or fear the reaction of the cartel if they should adjust their capital equipment to the current cartel price allows the cartel to charge a higher price than otherwise would have been the case. If the cartel should fall apart, then the other market participants might change their expected price only slightly, which implies further "punishment" for the cartel members in terms of a smaller market share than would have been the case in a deterministic and competitive world.

✳ *Chapter Six*

Analytical Units, Data, and Framework of the Study

The primary objective of this model is to establish the implications of various exporter strategies for the future price of oil in the international market. Because of the uncertainty associated with most econometric and engineering studies of the price responsiveness of oil/energy markets, the model was also designed to assess the sensitivity of the future price of oil to various coefficient or parameter values, however estimated. The uncertainty surrounding the traditional studies of the oil/energy markets results from the fact that the present level of prices is far outside the range of previous experience. Rather than allocating considerable time to determining point estimates of the various coefficients, a model structure has been designed in which the implications of the whole range of likely values for a given coefficient can be assessed.

To simplify the presentation of the model and its features, a set of coefficients estimated on the basis of OECD and FEA projections [6,8,10], as well as on the basis of the projections of other private and public institutions, has been chosen as a base case. The base case coefficients are hence estimated from the projections rather than directly from historic observations. The studies by OECD and FEA were designed to trace out the implications for the world's energy markets of changes in the price of crude oil in the Persian Gulf, the cartel price, which is the focus of this study. By assuming that the margin between the Persian Gulf price and the price paid by the final consumer would stay constant in absolute terms at the 1972 level, the net change in the future price to the final consumer of oil/energy

73

resulting from a given change in the price of crude oil in the Persian Gulf could easily be calculated. Once the net change in the price to the final consumer was established, the resulting net change in quantity consumed could be established by applying the elasticity figures resulting from numerous studies of particular oil/energy markets. The OECD and FEA studies are based on "what-if-a-price-of-$X-would-prevail-up-to-1985" scenarios. Both studies disregard uncertainty. The scenarios cover, however, a sufficiently broad range to indicate the responsiveness of the world's oil/energy markets to changes in the price of crude oil in the Persian Gulf. The responsiveness of the world's oil/energy markets to the Persian Gulf price is the basis for picking the "most desirable" price scenarios from the exporters' point of view.

The constant mark-up assumption, or the assumption that the margin between the price in the Persian Gulf and the price to the final consumer will stay constant in absolute terms, is conservative. Even if the transportation costs and company profits have come down slightly, and refinery and marketing costs have increased only modestly, the level of taxation of consumption has changed drastically in most parts of the world since 1972 [4]. The coefficients estimated and the results given in the following do not include price-induced consumer-country energy policies and should therefore be regarded as indicating how markets would behave without consumer government interference.

Based on the forecasts of numerous public and private institutions, estimates of the potential price responsiveness of the competitive exporter fringe, the non-OPEC producers and exporters, have been made. The assumptions underlying these estimates are made explicit. The reasoning behind the selection of behaviorally heterogenous exporter subunits is also made explicit, as are the assumptions underlying the calculations of the various coefficients and parameters characterizing the exporter subunits.

THE CONSUMER REGIONS

In the bathtub model, the world is divided into four consumer regions: Western Europe, Japan, United States, and the rest of the world (not including USSR, Eastern Europe, and China). The linear relationships representing the consumer regions in the bathtub model were estimated on the basis of the forecasts made by OECD, FEA, and other public and private institutions [7,8,9,10,11]. The procedure followed by these institutions to predict the future consumption of energy is:

1. Extrapolate the amount of energy spent in a base year (1972 or 1973) at the assumed rate of growth of the economy to some target year (most often 1980 or 1985). That is, a "long-term income elasticity" of one is implicitly assumed.

2. Reduce the amount extrapolated to the target date by a percentage equal to an assumed "long-term demand elasticity" times the percentage change in the price of energy from the base year to the target year.

We may define the "long-term elasticity of demand" to be the *ceteris paribus*, ultimate percentage change in quantity consumed resulting from a given percentage change of price in a perfectly deterministic world. I have defined the quantity subject to change differently from step 2 above, or more precisely as the initial quantity rather than the target-year quantity at the initial price. This implies that the absolute number of units of reduction will be lower than what the OECD-FEA procedure indicates. Only in the case of no income effects (no exogenous growth) are the two procedures identical. By incorporating the income effect, the price effect, as well as the length and distribution of the adjustment lags, the level of energy consumption in the target year as well as the intermediate years can be estimated as an explicit function of the above-mentioned variables.

By defining the quantity subject to change as the initial rather than the target-year quantity at the initial price, the price responsiveness of demand in absolute terms in a linear framework is scaled down. If we assume that the initial quantity was X_0, that the initial price was P_0, that at the initial price consumption was growing at G percent a year, and that it has been estimated that at a new price P_T then the quantity in some target year, T year hence, will fall to X_T, then information of this estimate would imply the following price responsiveness, or price slope:

$$\frac{X_O(1 + G)^T - X_T}{P_T - P_O} \qquad (6.1)$$

When assuming the same price responsiveness in relative terms affecting the initial level of consumption only, that is, that initial consumption will fall to a level X_{PO} such that

$$\frac{X_{PO}}{X_O} = \frac{X_T}{X_O(1 + G)^T} \qquad (6.2)$$

or

$$X_{p0} = \frac{X_T}{(1+G)^T}$$

then the price-slope calculated from the absolute changes will be smaller

$$\frac{X_O - \dfrac{X_T}{(1+G)^T}}{P_T - P_O} \tag{6.3}$$

The price slope (6.1) is consequently $(1+G)^T$ greater than the price slope (6.3). By scaling down the price responsiveness of the studies by OECD and FEA in this way, the price responsiveness assumed in this study may be considered to be very modest. The procedure by which the OECD and FEA estimates was scaled down makes it simple to understand the price assumptions underlying the simulations reported in the following as well as to change the price-slope assumptions for further simulations in a consistent and readily understandable manner.

To avoid kinked demand functions and discontinuous marginal revenue functions in a model basically designed to calculate price paths over time consistent with different sets of objectives of the oligopolists dominating the international petroleum market, energy demand functions, market share functions, and functions of indigenous supplies were linearized over the $3-$9 price range (in 1972 dollars). The price refers to the price per barrel of oil F.O.B. the Persian Gulf.

A more detailed discussion of the data assumptions of this study is included in section B.1 of Appendix B.

THE EXPORTER FRINGE

The existing and potential producers and exporters of the world that are not assumed to join an international cartel agreement, and which may become important noncartel sources of supply, are included in a separate unit, and the potential level of supplies of these countries, if they would behave as price-takers, are listed in Table 6-1. Brunei-Malaysia and Mexico might conceivably join an international cartel agreement. I have assumed they will not join OPEC. Greece, India, and Zaire are minors in a world context. They are included, however,

Table 6-1. Production of Non-OPEC Exporters MMB/D

	1973/74	1980		1985	
		$3	$9	$3	$9
United Kingdom	0.002	2.4	2.8	3.5	4
Norway	0.032	1.4	1.6	2.2	2.5
Canada	1.798	2	2.2	1.8	2.3
China[a]	0.090	2	2.6	2	2.6
USSR & Eastern Europe[a]	0.740	0.8	1	1	1.4
Brunei-Malaysia	0.320	0.7	1	0.7	1
Greece	0	0.06	0.06	0.18	0.18
Brazil	0.169	0.75	1	0.75	1
Mexico	0.465	1.5	2	1.5	2
India	0.148	0.2	0.2	0.2	0.2
Zaire	0	0.025	0.025	0.025	0.025
	3.764	11.84	14.49	13.86	17.21

[a]Includes net exports only. China produced 0.860 MMB/D in 1973. USSR and Eastern Europe produced 8.389 MMB/D.

Sources: *Petroleum Economist, Petroleum Encyclopedia 1974, Oil and Gas Journal,* various public and private institutions.

because their potential may be significant. The sensitivity of the future price of oil to assumptions about the level of supply from these sources should therefore be assessed.

The potential of the non-OPEC exporters as summarized in Table 6-1 was estimated in early 1975. Since then the potential of Canada has been scaled down following the increase in taxation and the phase-out of exports to the U.S. The export potential of China appears smaller today because of higher domestic consumption than originally anticipated. In the North Sea, in the British and the Norwegian sector, the rate of development has been slower than expected owing to higher taxes, uncertainty with respect to the two governments' policies, as well as technical problems in the hostile environment. Brazil has also revised its potential downwards. The net effect of the above-mentioned revisions[1] is a $9-1980 estimate of 9.5 million barrels a day (MMB/D), or 5 MMB/D less than in Table 6-1. The estimates of Table 6-1 may therefore be considered somewhat optimistic, even if they are well within the range of what might actually happen over the years to come. In the analysis reported in Chapter Seven, Table 6-1 represents the base case as far as the competitive fringe is concerned.

1. *The Oil and Gas Journal,* 1 December 1975, p. 89.

OPEC SUBUNITS

The composition of the exporter fringe seems plausible given the present configuration of OPEC. The membership of OPEC is, however, very heterogenous, and depending on the set of dimensions chosen to characterize the individual member countries, very different "homogenous" units may be defined [1,2,3].

To keep the complexity of a model representing a cartel-dominated market at a minimum, and to narrow down the range of possible market solutions, the number of explicitly colluding oligopolists should be kept at a minimum. In the following an attempt is made to collapse OPEC into a limited number of behaviorally homogenous subunits.

Using those factors that economic theory tells us are the most significant when explaining economic behavior, and on the basis of the observed behavior of the OPEC membership during 1973 and 1974, a number of dimensions have been defined. The positioning of each OPEC member along these dimensions has been used to select subunits within OPEC so that the homogeneity within each unit is greater than the homogeneity between the units. The individual OPEC countries are characterized along these dimensions in Table 6-2. It is apparent from the table that a number of different subcoalitions might be formed among these countries, depending on the dimension considered.

Columns one and two, "End-1974 Reserves" and "1974 Rate of Depletion," indicate the economic power of the various countries. A low rate of depletion indicates a high potential for expansion. A low rate of depletion and a high reserve level imply that the potential expansion of output could be substantial relative to the size of the market, and also that a smaller expansion could be implemented at a modest cost. The ability to expand output at a low cost is a measure of the economic power of a producer country. In addition to the production history of the country, the 1974 level of depletion indicates also the bargaining ability and power of the country as well as the relative emphasis on current as opposed to future revenues. Columns three and four, "1973-74 Percent Change in Rate of Depletion" and "1973-74 Percent Change in Production," indicate a country's willingness to produce below the potential to support the price level, the responsiveness of a country's reserve level to an increase in the price of oil, as well as the accommodations that will probably have to be made to keep the cartel together. Column five, "1974 Production Over End-1974 Capacity" is a proxy both for the short-term market power of the individual OPEC members and for

Table 6-2. Selection of OPEC Subunits

	End-1974[a] Reserves	1974 Rate[b] of Depletion	1973-74 Change in Rate of Depletion	1973-74 Percent Change in Production	1974 Production Over End 1974 Capacity	1974 Oil Income[c] Per Capita $	Embargo[d] Behavior	Location[e]
Saudi Arabia	173,100	0.018	-9.25	11.6	0.75	3600	E	PG
Kuwait	81,400	0.011	-24.696	-15.7	0.75	11434	E	PG
UAE Qatar	45,900	0.017	-29.817	5.24	0.73	6203	E	PG
Libya	26,600	0.021	-30.857	-28.1	0.64	4193	E	N-PG
Iran	66,000	0.033	-5.76	3.3	0.93	694	N-E	PG
Venezuela	15,000	0.002	-16.348	-11.6	0.92	1149	—	N-PG
Algeria	7,700	0.049	-9.395	-5.7	0.90	308	E	N-PG
Ecuador	2,500	0.022	-45.000	-29.8	0.48	98	N-E	N-PG
Indonesia	15,000	0.034	-25.202	4	0.84	35	N-E	N-PG
Iraq	35,000	0.020	-12.503	-2	0.76	805	—	N-PG,PG
Nigeria	20,900	0.039	4.483	9.7	0.95	124	—	N-PG
Gabon	1,800	0.036	-1.515	17.2	0.89	1300	N-E	N-PG
						415A		
						1198B		
Total	490,900	0.023	-14.818	-0.6	0.81			

aEnd-1974 reserves in millions of barrels. Source: *Oil and Gas Journal.* December 30, 1974.

bAnnual production divided by end-1974 reserves.

cAssuming the export price is the relevant value of domestic consumption and applying the population figures from the International Petroleum Encyclopedia 1974.

  A. OPEC average
  B. Average without Nigeria and Indonesia

dThe countries that actively participated in the 1973 embargo have been assigned an E in this column. The countries that significantly increased production during the embargo have been assigned a N-E in this column; and the countries that were close to neutral have not been assigned anything.

ePG means that the country has its most significant export terminals in the Persian Gulf. N-PG means that important export terminals are also located elsewhere.

the willingness and the ability of the countries to restrict current output, as well as for the bargaining ability and bargaining power of the member countries.

A number of dimensions may be defined to characterize the financial and political aspects of each country [2,3,5]. For the purpose of this study, however, the financial and political aspects are summarized in two dimensions only. Column six, "1974 Oil Income Per Capita" is an indicator of the financial position of each country. Higher per capita oil income signifies a lower need for present versus future revenues, a lower discount factor, higher financial reserves vis-à-vis import requirements, and a greater ability to survive production cutbacks. As the October 1973 embargo is the only event affecting the international petroleum market that clearly involved political considerations, the behavior of the OPEC membership during the embargo is used as a proxy for the political attitude of each country and for the political homogeneity of OPEC as a whole. Column seven, "Embargo Behavior," is an index of the producer countries' willingness to conform to joint political actions when there are substantial economic gains to be obtained along with the political gains, but when the distribution of the economic gains may be influenced by the behavior of the individual producer countries. The last column, "Location," indicates the ease with which the price/production policies of the membership can be supervised. The more different the characteristics and the location of a country's crude is from the marker crude, the more difficult it is to supervise its policies and to detect deviations from the joint policies, and the stronger is the incentive for the country to cheat. "Location" is a proxy for the ease with which the quality and transportation differentials as well as the credit terms of a country can be supervised.

From Table 6-2 it may be argued that OPEC consists basically of three different member categories. A first category consists of countries that could expand output substantially relative to the size of the international petroleum market but know that the market could accommodate their increased output only at a lower price. This category constitutes the "hard core" of the cartel. A second category consists of countries that are presently producing close to their potential, and that have a strong need for current income. The members of this category are the "price-pushers" within the cartel. They want to continue their present rate of production and prefer a higher to a lower price. A third category includes the countries that have smaller reserves than the members of the cartel core, have a strong need for current income, but are producing at a lower rate of depletion than the "price-pushers." The members of this category, the "expansionist fringe," are small relative to the market and would like to get a somewhat larger share of the market without having to reduce the price. That is, the expansionist fringe would like the other members to accommodate them with a higher market share at the expense of these other members.

The hard core of the cartel, which was denoted unit three, U_3, consists of the countries that have the largest recoverable reserves of oil, produce at the lowest rate of depletion, have the highest level of excess capacity, have the highest financial surplus level, and have demonstrated an ability to work together within a colluding agreement. The OPEC countries that fit this characterization are Saudi Arabia, Kuwait, UAE and Qatar, and Libya.

The price-pushers, which were denoted unit two, U_2, consists of the countries that produce at the highest level of depletion, have the lowest level of excess capacity, and have a strong need for current income. Iran, Venezuela, and Algeria may be included in this category.

The expansionist fringe, or unit one, U_1, consists of the countries that produce at an intermediate rate of depletion, expanded production the faster in 1973-74, have a high need for current income, did not participate in the embargo, and do not have their major export terminals in the Persian Gulf. Indonesia and Nigeria fit the characteristics of the expansionist fringe. Iraq has export terminals both at the Gulf and the eastern Mediterranean. It participated in the embargo, but with a show of independence. It is counted in this group, but obviously does not fit as well.

Ecuador and Gabon are not very significant in this context. The strong expansion of production in Gabon in 1973-74 made Gabon a candidate for the expansionist fringe, U_1. Ecuador's level of excess capacity was a decisive factor when including Ecuador among the price-pushers, U_2. Gabon's ability to further expand production may be limited, and Ecuador's level of excess capacity was not due to voluntary output reductions only, which implies that the two countries could both be accommodated elsewhere. The results of any analysis will not be significantly affected by such a change in status.

Even if Libya has relatively small recoverable reserves, the fact that Libya was the price leader of OPEC from 1970 to 1973, participated wholeheartedly in the embargo, and has also absorbed a significant share of the drop in the demand for imports made Libya a candidate for the cartel core.

There is still room for considerable expansion in Iran; the level of recoverable reserves is substantial, and the present rate of depletion is in the intermediate range. Iran did, however, expand production significantly during the embargo and was producing close to capacity in 1974 [13]. Iran's need for current income also makes Iran less likely to play the role of the residual supplier with the cartel core. The size of Iran's supplies vis-à-vis the international petroleum market makes it unlikely that Iran would expect other OPEC members to accommodate increased Iranian supplies, which leaves the price-pusher role as the most likely for Iran to play in the future as in the recent past.

Iraq's low rate of depletion, high need for current income, independent participation in the embargo, and Mediterranean export terminals, as well as the suspicion that recoverable reserves in Iraq are twice as high as those indicated in Table 6-2, are the factors that determined Iraq's membership in the expansionist fringe. It is considered unlikely that Iraq will be willing to keep production at the present low level.

For the purpose of simulating the "simple" model version, the model having one exporter group only, an "optimistic" competitive supply function was estimated for each of the OPEC subunits and these were then aggregated to serve as supply functions for the relevant coalitions of subunits. These supply estimates are considered optimistic because they assume that a supply level of one-tenth of the 1974 proven recoverable reserves could be sustained if the exporters behaved as price-takers and the price was $9 (in 1972 dollars). A reserves-to-production ratio of ten may be considered optimistic, but the assumption that the level of recoverable reserves will not continue to increase as a result of the fourfold increase in the price of crude oil may be considered pessimistic, as may also the assumption that the end 1974 production capacity is a proxy for the $3 competitive supply level. The optimistic case is therefore rather modestly optimistic, even if it is more optimistic than the supply assumptions underlying the simulations of the cartel version, namely that a reserve-to-production ratio of only twenty would be sustainable at a price of $9.

The coefficient and parameter values of the cartel units are discussed in sections B.1 and B.2 of Appendix B. There is a great deal of uncertainty associated with the guesstimated coefficient and parameter values. Our limited experience with high-priced oil and with the implementation of ambitious industrialization programs in the oil-exporting countries makes a discussion based on explicitly stated assumptions the most ambitious analysis of what will happen in the international petroleum market that can realistically be undertaken.

✳ *Chapter Seven*

Results of Bathtub Model
Simulations

To illustrate the nature of the model described in Chapter Five and Appendix A, and how this model combines aspects of formal modelling with informal story-telling to identify the likely evolution of price and trade pattern over time in a particular oligopolistic market, a set of stories, or of simulations of the simple version and of the cartel version of the model, is presented. The stories were constructed from cartel theory, from the empirical evidence on previous commodity cartels, and from the special characteristics of the individual exporters.

THE SIMPLE VERSION

In the simple version of the bathtub model the world is represented by four price-taking importer regions—Western Europe, the United States, Japan, and the rest of the world (not including the USSR, Eastern Europe, and China)—and one exporter unit. The simple version was designed to assess the significance of possible coalitions of exporter countries. The composition of the exporter coalitions can be changed by transferring countries from the exporter unit to the unit representing the rest of the world's indigenous suppliers, or vice versa. This version is labeled "simple" because all the exporters in the exporter unit are assumed to follow the same market strategy as opposed to the cartel version, in which the exporters may follow different market strategies.

The linear relationships of this simple version as described in Chapter Five were simulated with the coefficient and parameter values of Tables B-2, B-3, 6-1, and B-5. The simple version is hence simulated assuming the optimistic competitive supply response from the OPEC countries mentioned in the text to Table B-5. The optimistic competitive supply response implies that marginal production costs of the OPEC countries under the various monopoly scenarios of the simple model are assumed to be lower than in the case of the pessimistic supply response underlying the cartel model simulations.

In the optimistic case it was assumed that if the price of crude oil had increased to $9 in a competitive world, then the Persian Gulf producers could have supplied a quantity equal to one-tenth of the end-1974 proven recoverable reserves on an annual basis. In the "pessimistic" case, however, it was assumed that a $9 competitive price would sustain a level of production of only one-twentieth of the end-1974 proven recoverable reserves, once production equipment had been adjusted to that price. That the supply response is smaller in the pessimistic case reflects the implicit assumption that marginal production costs would rise faster in that case. Marginal production costs in both the optimistic and the pessimistic cases, constructed such that at the 1974 level of production marginal production costs in the model are $2.50, are way above the marginal cost figures reported elsewhere ($0.10-$0.70). The reason is that marginal production costs in this study were estimated or deduced from a hypothesis of what the supply response of the various producer countries would be if they would leave OPEC to behave as price-takers rather than directly from figures on production costs. The fact that marginal production cost is the inverse of the competitive supply function implies that if the hypothesized competitive supply response to a significant price increase is small, then marginal production costs rise fast, which implies that the slope of the linear marginal cost relationship assumed here will be steep.

We are concerned with constructing a competitive supply-marginal cost relationship that is a plausible approximation to the competitive supply curve in a competitive world and to the marginal cost curve in a monopolistic world, because we are concerned with the implications of both modes of behavior. The competitive supply-marginal cost relationship is common to both modes of behavior.

The pessimistic case was constructed because it was hypothesized that it would be more realistic than the optimistic case if some OPEC members should leave for the price-taking fringe. That is, the optimistic case implies an "unreasonably" large supply response in the case that an OPEC member would leave OPEC. The pessimistic case is a scaled-down version of the optimistic case. As it would be simple to draw conclusions about the impact of the two cases on the results from both model versions, and to avoid duplication of results, the two models were simulated assuming a different case for the purpose of demonstrating the nature of the models and their results.

Three different kinds of scenarios generated through the simple model version are reported below. To compare the results of this study to those of OECD, a number of scenarios were run to quantify the effect of the assumptions of this study as opposed to those of OECD.[1] These scenarios are summarized below.

The second set of scenarios was designed to trace out the price/quantity implications of various compositions of OPEC, ranging from OPEC falling apart to OPEC as a whole behaving as a monopolist.

The third set of scenarios traces the price/quantity implications of uncertainty as defined in Chapter Five.

The Bathtub Compared to the
OECD Predictions

The price responsiveness of oil imports assumed in this study is a scaled-down version of the price responsiveness estimated by OECD and FEA. The economic growth assumptions of OECD and FEA are far more optimistic than those of this study, as more recent projections taking account of the 1974-75 recession could be incorporated in this study. Use and output of oil by centrally planned economies are assumed to net out by OECD, which is a more conservative assumption than the net exports from these countries assumed in this study (Table 6-1). At a $9 price (1972 dollars) Table 6-1 projects a level of net exports from centrally planned economies of 3.6 MMB/D in 1980 and 5 MMB/D in 1985.

The more modest price responsiveness assumed in this study implies, *ceteris paribus*, an additional 9 MMB/D in 1980 and an additional 18.1 MMB/D in 1985 over and above the OECD projections. At $9 the more modest growth assumptions of this study imply, however, a projected reduction in imports of oil compared to the OECD case of 7.4 MMB/D in 1980 and of 10.64 MMB/D in 1985.

The net effect of the different assumptions is summarized in Table 7-1. The more modest price responsiveness of this study as opposed to that of OECD, results in additional demand for OPEC oil of 9 MMB/D in 1980 and 18.1 MMB/D in 1985. USSR and China are here assumed to export 3.6 MMB/D in 1980 and 5 MMB/D in 1985 as opposed to OECD's assumption of no exports from these countries.

1. OECD, *Energy Prospects to 1985* (OECD, 1974).

Table 7-1.   The Bathtub vs. the OECD Projections OPEC Production in Million Barrels a Day at a $9 Constant Price
(1972 Dollars)

	1980	*1985*	*1990*
OECD[a]	25.4	27.2	—
Bathtub	23.4	29.7	50.5

[a]OECD: *Energy Prospects to 1985* (OECD, 1974).

The more modest growth assumptions of this study results in a lower demand for OPEC oil, 7.4 MMB/D less in 1980 and 10.6 MMB/D less in 1985 than the demand resulting from the OECD growth assumptions.

The 1973-90 OPEC production path of this study at a $9 constant price is indicated in Figure 7-1.

OPEC Breakdown
From Table 6-2 the end-1974 production capacity of the OPEC countries appears as approximately 38 MMB/D. The $9 scenario of Figure 7-1 ($9 in 1972 dollars, which would be about $11.70 in 1975 dollars) indicates that OPEC's daily production might fall as low as 25 MMB/D on an average annual basis as early as 1979. Even if OPEC managed to keep the cartel discipline of April 1975, in

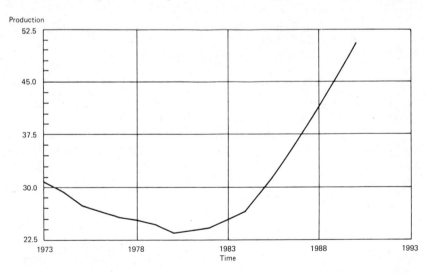

Figure 7-1. OPEC Production at $9 Constant Price 1973-90 (Production in Million Barrels Per Day, MMB/D)

which month the OPEC countries produced at a rate of 25,888 MMB/D[2] only, no international commodity cartel has managed to live with such a huge excess capacity (see Chapter 3). It may therefore be prudent to question OPEC's ability to live with an excess capacity of about 35 percent of the presently existing total productive capacity in the period from 1974 to 1985. That the OPEC countries are still adding to the level of productive capacity will increase the strain on cartel discipline even further.

If we assume that OPEC as a cartel started in 1973, then 1979 also coincides with the expiration of the period over which the average efficient cartel was able to survive. A scenario was therefore constructed in which the cartel manages to keep a constant price in real terms from 1974 to 1979 and then collapses in 1979, and the static competitive price emerges in the same year. The price/production path of this scenario is depicted in Figure 7-2.

The long adjustment period of demand accounts for the slow pickup of OPEC sales and production. The conservative assumptions underlying the OPEC competitive supply functions account for the apparently high price level that emerges when the cartel breaks down. The 1979 price is the long-term marginal production cost of supplying the 1979 level of production of 26.16 MMB/D within the OPEC countries. The price level in 1979 is equal to the long-term static competitive price level, and the fact that marginal production

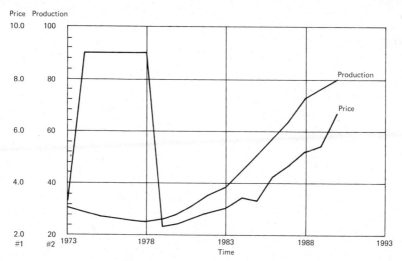

Figure 7-2. Price and Production Pattern in the Case of OPEC Breaking Down in 1979 (MMB/D, $1972)

---

2. *Petroleum Intelligence Weekly*, 26 March 1975, p. 11.

costs when producing from an already developed field are smaller than the marginal production costs when additional capacity has to be developed to produce an additional unit is disregarded. The cost functions of the model include both production and development costs independent of the level of productive capacity, which implies that the price level in the case of OPEC breakdown might fall even farther to the marginal production costs of producing from existing facilities.

For a number of reasons, some of which are pointed out in the cartel model scenarios, a cartel breakdown could be extremely costly for the OPEC membership. It is therefore likely that some effort will be made to find a more favorable solution by all or some fraction of the OPEC membership.

OPEC Monopoly Prices

Should OPEC follow a joint-profit-maximizing strategy, the OPEC- "monopoly" price/quantity path should emerge. In Figure 7-3 the price/quantity path of an OPEC monopoly strategy is indicated. The monopoly price is the higher of the static monopoly price and the exhaustible-resource monopoly price defined in Chapter Five. Even when the ultimate level of recoverable reserves is assumed to be equal to the end-1974 level of recoverable reserves only, the backstop technology price is assumed to be $15 in 1972 dollars (or about $19.5 in 1975 dollars), and the appropriate discount factor 10

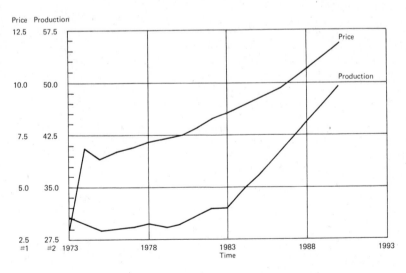

Figure 7-3. OPEC Monopoly Price and Production ($-1972, MMB/D)

percent, then the exhaustible-resource monopoly price is smaller than the static monopoly price all the way to 1986. In 1987 the exhaustible-resource monopoly price is $0.08 higher, and in 1990 $0.61 higher than the static monopoly price. That is, the opportunity cost of oil as measured by the net present value equivalent of the backstop price is smaller than marginal production costs in the period from 1974 to 1986.

The OPEC monopoly price, the joint-profit-maximizing price resulting from perfect collusion among the OPEC members, is far below the present price level. The reason for this discrepancy may be that the OPEC members have a different notion of what is in their best interest, that the bargaining power of some individual member is so strong as to divert the joint price away from the joint-profit-maximizing price to a level consistent with profit-maximizing behavior on the part of some individual member only, or some assumption may be mistaken.

It may also be noted that OPEC's present price appears to be consistent with exploitation of the short-term elasticity of demand, and that the price under such an hypothesis would be expected to come down to the longer-term monopoly level. The bathtub model as it stands today is not, however, designed to identify these shorter-term aspects of a recently monopolized market.

One purpose of the cartel model is to identify the desired price and production paths of the individual OPEC members and thus also to reveal the extent to which the desired market solution of some OPEC member dominates OPEC policies. However, an initial impression of the "desires" of the individual OPEC members may be obtained also from the simple model, as demonstrated below.

The heterogeneity of the OPEC countries implies that each member may define a different "OPEC-monopoly price" dependent on each country's anticipated share of OPEC production and on the costs of producing that share of the total. By disregarding the regional relationships between producer and consumer countries, and by constructing marginal-cost functions for each OPEC member consistent with Table B-5, each OPEC country's OPEC-monopoly price may be identified in the simple bathtub model. The residual demand facing a country j in period t, D_j^t, as a member of OPEC, may simply be represented as the anticipated share of total OPEC production, W_j^t.

$$D_j^t = W_j^t \times D^t \tag{7.1}$$

In Table 7-2 the desired 1975 OPEC monopoly price is indicated

Table 7-2. Desired 1975 OPEC Prices in 1972 and 1975 Dollars

	Algeria	*Venezuela*	*OPEC*
1975 dollars	9.3	10.86	8.27
1972 dollars	7.15	8.35	6.37

for Algeria, Venezuela, and for OPEC as a whole. Algeria and Venezuela were picked because they had the highest rate of depletion or had the lowest reserves-to-production ratio (Table 6-2). A low reserves-to-production ratio implies that a country is closer to its own resource limit, and therefore may be interested in switching to the backstop price at an earlier date. It was assumed that Algeria and Venezuela would be assigned a production quota equal to the two countries' share of OPEC's 1973 production.

That the current price is so close to Venezuela's desired OPEC price tends to strengthen Venezuela's image as the expert among the oil-producers and as an internal price-pusher. The reason for Venezuela's higher desired price is that the end-1974 level of recoverable reserves, which is used as a proxy for ultimate recoverable reserves, is small compared to current production, so that the opportunity cost of oil in Venezuela as measured by the net present value of the backstop price is substantial. The net present value equivalent of the backstop technology price is significantly higher than static production costs in the Venezuelan case.

There is reason to doubt some of the OPEC members' willingness to adhere to jointly determined quota policies even if OPEC were ever to be able to design and implement a prorationing system. If some OPEC countries are expected to increase production to a level such that marginal production costs are equal to the OPEC-determined monopoly price rather than stick to some "fair" market share, then these OPEC members may be considered as price-takers, and the relevant monopoly unit may be defined to be OPEC minus these price-takers, some subunit of OPEC. The simple bathtub model was consequently simulated assuming that two OPEC subunits would play the monopoly role.

In Figure 7-4, the monopoly price path resulting from throwing Indonesia, Nigeria, Iraq, and Gabon into the indigenous supply function of the "rest of the world," and letting the remaining OPEC countries follow a monopoly strategy, is plotted. The price path resulting from letting only Saudi Arabia, Kuwait, UAE, and Libya play the monopoly game, when everybody else is a price-taker, is also plotted along with the joint OPEC monopoly price.

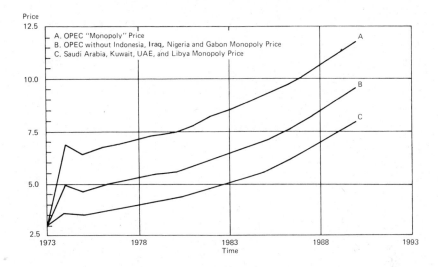

Figure 7-4. OPEC and OPEC Subunit Monopoly Prices ($-1972)

All OPEC countries have reduced production somewhat, but not everybody has reduced output to the same extent, which implies that the optimal price from the dominant unit's point of view should recognize price-taker tendencies and would probably lie somewhere in between the price paths of Figure 7-4.

These simulations indicate that the present price is consistent with a hypothesis of price-pusher influence. As the economic sophistication of the dominant unit (Saudi Arabia, Kuwait, UAE, and Libya) increases over time, a shift downwards in price may be expected, even if OPEC should survive as a colluding unit.

The Effect of Uncertainty

The uncertainty associated with the future price of oil may make the level of excess capacity associated with a market price of $9 smaller than otherwise would have been the case. A number of private and public institutions have predicted a drop in the price of oil in the late seventies or early eighties owing to the anticipated flow of oil from non-OPEC producers and exporters as well as the emergence of some intracartel competition. The effect of expected prices different from actual prices is demonstrated in the bathtub model by constructing an expected price, $E(P)$, different from the actual price, P. In this scenario it is assumed that the expected price is a weighted average of the actual price and a $6 price.

$$D^t(P) = 0.5 \times 6 + 0.5 + p^t.$$

The actual price is kept constant at $9. The market will consequently adjust to an expected price of $7.50. By comparing this scenario with the $9 scenario in which actual and expected prices coincided, as done in Figure 7-5, the quantity of world imports is kept at a significantly higher level over the time horizon of the model in the case of the expected price being lower than the actual price. The difference in the 1974-85 period is possibly large enough to save the cartel from collapsing because of an inability to allocate the necessary reductions in output.

If OPEC should try to exploit the fact that the expected price is different from the actual price, the price charged in each period would be the higher path in Figure 7-6, whereas the expected price would be the lower path of Figure 7-6. The expected price is as defined above, halfway between $6 and the actual price. The expected price in the "uncertainty monopoly" case is equal to the actual price in the "normal" monopoly case. Under the monopoly strategies defined in Chapter Five, the monopolist is assumed to maximize revenues with respect to the expected price. The monopolist is also assumed to have perfect information with respect to the probability distribution determining the expected price.

Exploiting the uncertainty as represented by the weights of the

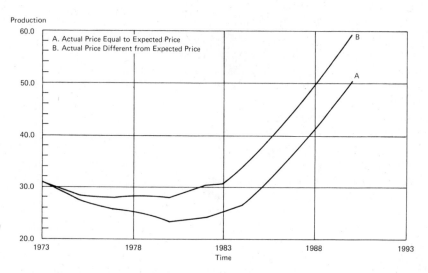

Figure 7-5. OPEC Production at a $9 Actual Price and Different Expected Prices ($-1972, MMB/D)

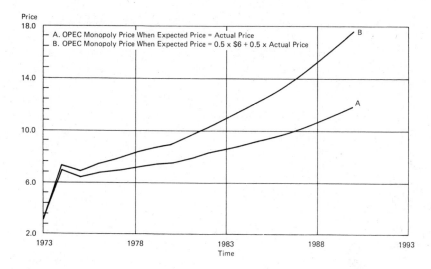

Figure 7-6. OPEC Monopoly Price Under Different Assumptions about Price Expectations

expected price formula makes it possible for the cartel to profitably charge a substantially higher price than a stable monopoly would have been able to do.

In Table 7-3 the simple model scenarios are summarized in 1975 dollars (assuming a 1972-75 rate of inflation of 30 percent) and in million barrels a day of OPEC production. It is apparent that over

Table 7-3. Summary of Simple Model Scenarios (Prices in 1975 Dollars-Production in MMB/D)

	1980		*1985*		*1990*	
	Price	*OPEC Production*	*Price*	*OPEC Production*	*Price*	*OPEC Production*
$9 Path (1972 dollars)	11.7	23.4	11.7	29.7	11.7	50.5
OPEC Breakdown 1979	3.13	28.2	4.93	50.4	7.33	79.6
OPEC Monopoly-Pricing	9.65	29.6	11.92	36.6	15.37	49.4
OPEC Without Fringe	7.16	26.3[a]	9.07	34.9[a]	12.44	46.8[a]
Cartel Core Alone	5.53	19.5[b]	7.19	30.2[b]	10.37	41.1[b]
$9 and Uncertainty	11.7	27.9	11.7	37.3	11.7	59.5
OPEC and Uncertainty	11.48	29.6	16.02	36.6	22.75	49.4

[a]Production of OPEC without Indonesia, Iraq, Nigeria, and Gabon.
[b]Production of Saudi Arabia, Kuwait, UAE, and Libya.

the next ten years the likely price range is substantial even if the likely production and hence also the likely exports range is much narrower due to the inertia of the oil/energy markets. The uncertainty scenarios indicate that the present price of about $10.46 (in 1975 dollars) is too high in a joint OPEC setting. (The 1975 "OPEC-uncertainty" price in 1975 dollars is $8.9.) The tendencies of some OPEC members to be unwilling to reduce their output with the market do, however, indicate that a price somewhere in between the "OPEC and Uncertainty" price and the "OPEC without Fringe" price would be more favorable to the larger OPEC members.

THE CARTEL VERSION

From the simulations of the simple bathtub model, or from the stories that were being told by simulating that model, some of the critical aspects of the future international petroleum market emerge. A critical aspect is the extent to which the exporting countries can agree on joint, as opposed to individual, strategies. The composition of the colluding group of exporters directly determines the level of the monopoly price. The uncertainty associated with the composition of and the stability of the most likely colluding groups determine to a large extent the response of the world's energy markets to the pricing strategies that may be pursued by the various colluding groups and thereby also the joint-profit-maximizing price as perceived by the exporters.

To focus on the likely composition and the stability of collusive exporter groups, it is necessary to consider the collusive agreements from the point of view of the individual exporters. The stability of a collusive agreement depends on the desirability of the agreement to the individual members. A model that can tell stories about the desirability of various collusive arrangements to the individual participants is therefore needed to consider the stability question. The cartel version of the bathtub model is designed to tell stories about the possible reaction of individual exporter units to various collusive arrangements.

In the simple model version the world is collapsed into two groups of market participants, the importers and the exporters. All the members of each group follow the same market strategy, even if the composition of the two groups can be changed to demonstrate the significance of likely exporter coalitions. The forces working on the coalitions, or the incentives to form coalitions, cannot be explicitly discussed within the framework of the simple model. By dividing the world into one importer group and four exporter units, the level of conflict within various coalitions of the exporter units may be explicitly discussed. The level of conflict depends on how close the actual joint market strategy of the colluding group is to the joint strategy considered most favorable by the individual exporters. The level of tolerance for internal conflict depends on the attractiveness of the individual market strategies open to the members relative to the outcome of the actual joint market strategy.

In the cartel model some joint and individual market strategies are compared from the point of view of the individual exporters. Criteria have been defined that indicate the attractiveness of various market strategies to the exporters. By allowing noncolluding exporters to follow noncompetitive market strategies, the contribution of such individual noncompetitive strategies to the maintenance of a noncompetitive price level over time may be assessed. The composition of the individual exporter units and the reasoning behind the selection of these units are indicated in Chapter Six.

The first of the cartel stories focuses on the implications of the constant $9 path (1972 dollars) to the OPEC membership. The financial and excess capacity situation of each member is of primary interest.

There are, however, a number of alternatives open to OPEC to improve upon their joint $9 arrangement. Some stories are told that indicate the kind of specific alternatives the various individual members of OPEC might suggest. The implications of these alternatives for the OPEC membership, the acceptability of the alternatives, as well as the implications for the international petroleum market are also discussed.

Should the proposed alternatives be unacceptable to some members, they might leave OPEC. The consequence of such action is discussed. The financial advantages of being a member of an "imperfect" OPEC from the point of view of some individual exporter may be considered a proper compensation for the kind of compromise the exporter would have to agree to as an OPEC member. The incentive to reestablish OPEC in case of some OPEC cracks is also discussed.

The fact that non-OPEC exporters may also follow nonconventional market strategies is explored. The strengthening of OPEC resulting from such behavior as well as from the perceived uncertainty with respect to the future price of oil/energy by the world's oil/energy consumers and producers is the focus of that story.

Historic Quotas
Cartel theory cannot give us the necessary and sufficient condi-

tions for a cartel-dominated market to remain stable over time. It is possible, however, to simulate a cartel-dominated market, to assume a stable price, and then to demonstrate what such a stable path would imply for each cartel member in terms of some critical factors. It is thereby possible to demonstrate the minimum level of discipline each cartel member must live with, and this may be used as a proxy for the necessary conditions for any stable price path to be observed.

The cartel model was therefore simulated to identify some of the implications of a constant price in real terms of $9 (in 1972 dollars or approximately $11.70 in 1975 dollars) and of cartel production being allocated over time as it was in 1973. This scenario is hence a breakdown of the implications of the $9 scenario discussed above for each of the exporter units, and thus analogous to the exercise made by FEA [FEA 1974] to indicate the kind of pressure that OPEC would have to sustain to keep a $9 price level (in 1972 dollars).

An important aspect of the pressure that will build up is the level of excess capacity with which the cartel will have to live. The present level of productive capacity, approximately 38 MMB/D, is far above the expected production in the years to come. As there is no system of prorationing of capacity within OPEC, and the present price level is far above the cost of installing additional capacity in the producer countries, the producers may be expected to increase their existing capacity. A number of individual and joint capacity strategies, however, may be designed by the producers. To indicate the potential level of capacity, and hence also of excess capacity, it was assumed that the exporters will develop a level of productive capacity equal to their supply potential in a $9 competitive world. The supply potential, or the competitive supply level given a $9 price, was determined assuming the coefficient and parameter values of Table B-6.

In 1973 the expansionist fringe produced 17.4 percent of total OPEC production, the price-pushers produced 35 percent of the total, and the cartel core produced the remaining 47.6 percent. Table 7-4 lists the actual income and the import requirements of the expansionist fringe and the price-pushers, and the excess capacity of the three individual cartel units. It is unlikely that the OPEC countries will develop a production capacity equal to the supply potential. The level of excess capacity as defined in Table 7-4 does, however, point to an important aspect of the strain to which OPEC might become subject.

The "actual income" consists of the income associated with oil production and with holding claims on (having liabilities to) foreign countries. The projected import requirements are simply a function

Table 7-4. Implications of a $9 Price and Historic Quotas
(Income in Billions of 1972 Dollars, Excess Capacity in MMB/D)

	Expansionist Fringe			Price-Pushers			Cartel Core
	Actual Income	*Import Requirements*	*Excess Capacity*	*Actual Income*	*Import Requirements*	*Excess Capacity*	*Excess Capacity*
1977	15.73	14.34	4.68	34.48	18.77	3.33	26.9
1978	15.35	17.21	5.7	34.32	22.52	3.93	33.51
1979	14.57	20.65	5.9	33.46	27.02	4.33	34.06
1980	13.61	24.78	6.09	32.38	32.43	4.73	34.60
1985	14.82	40	4.8	40.21	52.23	2.12	31.06
1990	8.9	64.27	1.19	20.7	84.11	0	16.03

of time. The coefficients of these time-dependent functions are listed in Table B-6, as is also the initial level of assets abroad for each of the cartel units. Nonpetroleum exports are disregarded, which implies that the projections made beyond 1985 are very inaccurate. It may take ten years before nonpetroleum exports become significant.

Table 7-4 is stated in constant 1972 dollars. It is implicitly assumed that the rate of inflation of oil prices will equal the rate of inflation of import prices to the OPEC countries, which may tend to bias the import requirements downwards and the level of accumulated assets abroad upward (and thus actual income upwards), as the higher rate of imports inflation may blow up the import requirements. According to *Middle East Economic Survey*, the import price index rose on the average for each OPEC country by 26.3 percent in 1974 (up from 10.3 percent in 1973 and 1.1 percent in 1972), or substantially faster than any worldwide index of inflation that might be used as a yardstick for indexing of oil prices.

Even under the optimistic assumptions of this scenario, by 1990, OPEC holdings abroad will on the average have reached the zero level again, even if there are substantial individual differences. As Table 7-4 shows, the expansionist fringe will run a deficit as early as 1978, and the price-pushers will run a deficit already in 1980. The cartel core will not run a deficit in the simulation period. It is apparent, however, that the cartel core will bear most of the burden of the excess capacity as defined above.

A number of stories can be told about why some cartel member would or would not like the $9 stable price and quota path described. Some of the stories we consider plausible in terms of what we might expect the individual cartel members to suggest as common policy and how the other cartel members might react to these suggestions will be explored. The stories are put together or constructed on the basis of cartel theory, the empirical evidence on international commodity cartels in general, as well as on the basis of the specific characteristics of the OPEC countries outlined in Chapter Six.

Financial Quotas

The expansionist fringe of OPEC, Indonesia, Iraq, Nigeria, and Gabon has a high need for income, and on an average per capita basis, the level of oil income is small both in absolute terms and compared to the rest of OPEC. These countries may therefore be reluctant to accept any reduction in output that would not instantaneously give higher export revenues, and they may become very hard-pressed economically as well as politically when the level of income falls short of their income requirements, which might happen in early 1978 (Table 7-4). A possible reaction, given their own financial situation as well as the financial surpluses of the cartel core, would be to propose a quota system based on the income requirements of the cartel members as defined in Chapter Five. Such a quota system would significantly increase the market share of the expansionist fringe and the price-pushers (Table 7-5). But even such a significant increase in market share would only marginally postpone the point in time when oil and financial income would fall short of the rapidly increasing income requirements of these countries. The income difference under the two scenarios is, however, substantial.

The cartel core is subsidizing the other cartel members under the financial quota scenario. The rate of depletion of the cartel core's reserves drops to a level 50 percent lower than in the historic quota scenario, even if the market share of the price-pushers drops fairly rapidly after 1985 and the market shares of the expansionist fringe decreases after 1986 due to capacity constraints and an expanding market. The number of periods the cartel core has to wait to sell lost current output makes the net present value of the additional future cash flow associated with selling "oil in the ground" so insignificant that the additional current income of the noncore members represents an income transfer, a side payment, from the cartel core to the other members approximately equal in magnitude to the net gain of the noncore members. The cartel core may be reluctant to grant a subsidy that significant and may react by proposing an alternative price-quota system.

Price-Pusher Dominance

If it is considered unfair to fiddle with the historic market shares, then the price might be used to increase the revenues of the cartel as long as the current price is below the monopoly price. By agreeing to a joint pricing strategy that is "optimal" from the "low-reserve

Table 7-5. Financial as Opposed to Historic Quotas

| | Expansionist Fringe | | Price-Pushers | |
	Historic Quota	Financial Quota	Historic Quota	Financial Quota
1978	0.174	0.316	0.35	0.414
1980	0.174	0.33	0.35	0.433
1985	0.174	0.33	0.35	0.423
1990	0.174	0.20	0.35	0.25

base," "high need for income" countries but too high for the low-cost, "extensive reserve base" countries, a side-payment system across time is in effect being implemented. If, therefore, the cartel core should object to a quota system based on income requirements, the noncore members might propose a pricing strategy that would contain such a time-dimensional side-payment system.

In this scenario it was assumed that the financial quota system would survive only for a year (1978), and that the OPEC pricing strategy desired by the price-pushers, Iran, Venezuela, Algeria, and Ecuador, would be implemented in 1979 along with the historic quotas. Even if the optimal 1979 price is below $9, the optimal price increases fast enough to significantly alleviate the financial problems of the noncore members (Table 7-6). The pricing strategy of this scenario produces more income for the noncore members than the financial quota system. The depletion rate of the cartel core's reserves is also substantially higher under the historic quota system than under the financial quota system, even if it is still low, which may make the price-pusher strategy more acceptable to the cartel core.

Cartel Core Dominance

It may also be assumed that the sophistication of the OPEC decision-makers, including those of the cartel core, will increase over time, and that such a radical proposal as the financial quota system might trigger an adjustment of OPEC policies to some economic realities. Such an adjustment might, of course, take place even without any prior radical proposals.

The economic realities are OPEC's monopolylike position and the economic power of the individual cartel members. OPEC may therefore choose to follow a joint-profit-maximizing pricing strategy and to allocate production and profits according to the "economic power" quotas (Chapter Five). An "economic power" quota system

Table 7-6.   Actual Income Under Price-Pusher Price and Historic Quotas as Opposed to $9 Price and Historic Quotas

	Expansionist Fringe		Price-Pushers		Market Price
	Price-Pusher Price Income	$9 Price Income	Price-Pusher Price Income	$9 Price Income	Price-Pusher Price ($1972)
1980	14.08	13.61	32.87	32.38	9.00
1985	17.28	14.82	46.63	40.21	11.95
1990	23.38	8.9	62.45	20.4	15.40

implies that each noncore cartel member is assigned a production quota such that the particular member makes at least as much profit as a cartel member as he would have made as an outside price-taker. The cartel thereby minimizes the compensation a cartel member will want to adhere to the policies of the cartel. Under this scenario, the cartel core, being the dominant producer, allocates the profits and keeps the residual rent.

Such a scenario was simulated and some of the implications are listed in Table 7-7. It appears that the historic quotas are not very different from the economic power quotas. The expansionist fringe would gain some, and the price-pushers would lose some compared to the historic quotas. The cartel core would consequently be able to increase production only marginally compared to the historic case.

Both the expansionist fringe and the cartel core are better off under this scenario than under the historic scenario. The expansionist fringe should be close to indifferent between the economic power scenario and the price-pusher scenario as indicated by the income columns of Tables 7-6 and 7-7. The "price-pushers" are, however, worse off in the latter than in the former case. The over-all implications of the economic power scenario would make it attractive to OPEC.

The OPEC monopoly price of Table 7-7 is higher than that of Table 7-3 because of the higher production costs assumed in the cartel model simulations.

OPEC could, however, do a little better by allocating production and profits as a multiplant monopolist (Chapter Five). Under such a scenario, production would be allocated in the most efficient way. That is, marginal production costs would be equal in all cartel units at the allocated level of production. It would thereby be possible to obtain an efficiency premium over and above the economic power quota system. Such a scenario is not presented here because of the bias of the linear relationships.

The marginal cost functions of the bathtub model are linear. The

Table 7-7. Economic Power Scenario

	Expansionist Fringe		Price-Pushers		
	Economic Power Quota	Actual Income	Economic Power Quota	Actual Income	Market Price ($1972)
1980	0.19	15.04	0.29	26.63	8.53
1985	0.18	19.44	0.25	31.78	10.65
1990	0.17	22.62	0.22	40.36	13.13

coefficients determining the location and slope of the marginal cost relationships were estimated on the basis of an assumed expansion. This implies that for the countries that could only marginally expand production under an assumed increase in the competitive price level, the marginal cost curve would be very steep. This implies again that in the same countries the marginal costs would be assumed to fall very rapidly if the countries should produce short of capacity. The effect of the linear marginal cost relationships is that countries having a steep marginal cost curve and producing short of capacity would be allocated a share of monopoly production that is biased upwards.

Linear marginal cost relationships also imply that average cost is equal to marginal costs at a production level 50 percent of the current production level, as can be seen by integrating the linear marginal cost relationship to find total costs and then by dividing by quantity to get average cost. In the case of a negative average cost when deducted from the linear marginal cost relationship, a kinked average cost curve was assumed to avoid the negativity problem. A flat constant average cost minimum was assumed up to the point where average costs consistent with the marginal cost relationship becomes positive. Under competitive supply assumptions the linear relationships imply that average cost is implicitly assumed to be half the competitive price. As the average cost of producing from existing capacity is only a fraction of the cost of adding new capacity, which is the relevant cost concept in a competitive world, the above-mentioned linear relationships will tend to bias downwards the economic power quotas. The profits of a cartel unit as an outside price-taker will be biased downwards.

These biases, even if they would be substantial in the multiplant allocation of production case, are not considered of sufficient importance to make the linear relationships unfit in a model intended to explore the implications of some plausible exporter strategies. Once the implications of some exporter strategies have been evaluated in this simplistic framework, more sophisticated functional relationships may be introduced. Because of the computational problems involved when formulating more sophisticated relationships, and the resulting dramatic increase in the costs of simulating a bathtub model, the introduction of more complex relationships is considered to be beyond the scope of this study.

OPEC Cracks

None of these cartel scenarios allow the expansionist fringe to earn enough on its oil production to cover the imports bill for the period beyond 1978. It may be politically unfeasible for the expansionist fringe to produce at a level short of full capacity when at the same time there is a shortage of funds. The other members of OPEC may recognize the fringe's inability to restrict production according to an OPEC-designed quota system, and choose a pricing strategy accordingly. That is, the other members of OPEC recognize the expansionist fringe's inability to behave differently from a price-taker. The price-pushers and the cartel core may consequently try to maximize joint profits under the assumption that the expansionist fringe will behave as a price-taker and produce at capacity. A cartel scenario was therefore constructed in which the expansionist fringe was assumed to follow a price-taker strategy from 1979 and beyond, and the production and profits of the monopoly unit, the price-pushers and the cartel core, were allocated according to the economic power quotas. The resulting market price in 1979 is $1.14 (in 1972 dollars) below the joint-profit-maximizing price of OPEC as a whole. The 1979 income of the price-pushers is approximately $27.1 billion in the joint OPEC case with economic power quotas, as opposed to only about $16.2 billion in this scenario. The dramatic drop in income results primarily from the expansionist fringe throwing its full potential capacity of 10 MMB/D into the market in that year. The level of excess capacity of the price-pushers is also high as a result of the expansionist fringe following a price-taker strategy. A likely response of the price-pushers may therefore be to follow the example of the expansionist fringe and produce at capacity as a price-taker. The immediate effect of such a move by both the fringe and the price-pushers would be very dramatic for the cartel core. To maintain its profit-maximizing price in 1979, the cartel core would have to restrict output to a mere 1.75 MMB/D in that year.

The dramatic effect on the production level of the cartel core might also result in a price war. A price war could even push the price close to a zero level.

Reestablishment of OPEC

In the industries whose structure is most favorable for collusive arrangements, there have been a number of attempts to reestablish a collusive agreement once an agreement has fallen apart. Chapter Three gives ample evidence on this aspect of "oligopolluted" industries. There have always been considerable short-term incentives to reestablish a noncompetitive price level, even if in the longer term the noncompetitive price level often induced more entry than anticipated, to the detriment of the colluding firms.

In the case of an OPEC breakdown there would be considerable incentive to reestablish a collusive arrangement. The difference between the income of the cartel core when maximizing profits alone and when reestablishing OPEC is more than $35 billion in 1981. The 1981 difference between the same two scenarios is more than $2 billion for the expansionist fringe and more than $6 billion for the price-pushers.

If a price-war price level rather than the long-term competitive equilibrium price level emerged following an OPEC breakdown, then the incentive for reestablishment would be even stronger. The emergence of a long-term competitive equilibrium price in case of OPEC breakdown is based on the assumption that if price should fall below this level, there would be a strong enough speculative demand for oil to force back the price to the long-term competitive level.

The confidence of the OPEC countries in running a collusive arrangement might get a serious crack in case of a breakdown. Lack of confidence might make future collusive arrangements even more unstable, and we would expect to observe a fluctuating price level due to a "breakdown-reestablishment-cycle." One such cycle is included as an illustration in Table 7-8, the market price "cycling" from $9 in 1978 to $3.94 in 1979 and up to $9.50 in 1981.

Income-Stabilizing Exporter Fringe and Uncertainty

Some of the governments of the countries we have assigned to the competitive exporter fringe have expressed concern over the sheer magnitude of their expected oil export revenues. High export revenues may, if they were to be absorbed immediately in the local economy, cause inflation only. Production ceilings have consequently been constructed to adjust future production to a level of income that the governments believe the local economies can absorb. The reasoning is thus along the lines of the income stabilization strategy (Chapter Five).

Table 7-8. Breakdown-Reestablishment Cycle

Year	Market Price ($72)
1974-78	9.00
1979	3.94
1980	4.87
1981	9.51
1985	10.68
1990	13.18

The implication of a competitive exporter fringe behaving as on a backward-bending supply curve is that the monopoly unit can disregard the exporter fringe and choose a monopoly price with respect to the demand for world imports, D^t, which is higher than the monopoly price with respect to world residual imports, RD^t.[3]

If we assume that the competitive fringe had anticipated an income level consistent with a $9 price and production at capacity, and that they would reduce production according to Equation (5.12) if the price moved above this level, then OPEC could raise price almost 20 percent higher in the second half of the breakdown-re-establishment cycle (Table 7-9) than otherwise would have been the case (Table 7-8).

The dramatic jump in demand for cartel output following the reestablishment of OPEC in 1981 is caused by the competitive fringe's reduction in output following the price increase in the same year. The income-stabilization strategy of the competitive fringe will greatly increase the attractiveness of a joint OPEC monopoly pricing strategy, increasing both the price and production level of the OPEC countries and, consequently, the income level of these countries.

If the noncartel market participants adjust their consumption and production equipment to an expected price level that is below the price currently charged by the cartel, then such behavior will have a price effect similar to that of an income-stabilizing exporter fringe. That is, the cartel can charge a higher price and will be able to produce a larger quantity than otherwise would have been the case.

A scenario was constructed that adds uncertainty, as defined in

Table 7-9. Income Stabilizing Fringe

Year	Market Price ($72)	OPEC Production MMB/D
1974-78	9.00	—
1979	3.67	24.76
1980	4.73	25.9
1981	12.02	32.72
1985	13.07	35.06
1990	15.52	39.87

3. By substituting Equation (5.12) for SE^t in Equation (5.15), and taking the first derivative of the revenue function associated with Equation (5.15), the implication above becomes obvious:

$$\frac{\partial(RD^t \times p^t)}{\partial p^t} = \frac{\partial(D^t - [I^t/p^t])p^t)}{\partial p^t} = \frac{\partial(D^t \times p^t)}{\partial p^t} - \frac{\partial I^t}{\partial p^t} = \frac{\partial(D^t \times p^t)}{\partial p^t} \quad (7.2)$$

Chapter Five, to the income-stabilizing case above. It is assumed that the expected price, $E(P)^t$, is a weighted average of the long-term competitive price, PC^t, the stable price path, $9, the backstop technology price, $16, and the joint OPEC price, $p4^t$, and the OPEC subunit prices of the price-pushers and the cartel core, $p5^t$, and of the cartel core alone, $p3^t$, as well as the current price, p^t, in the following way:

$$E(P)^t = 0.1 \cdot PC^t + 0.1 \cdot 9 + 0.1 \cdot 16 + 0.1 \cdot P4^t +$$
$$0.1 \cdot P5^t + 0.1 \cdot P3^t + 0.4 \cdot P^t$$

The expected price of this scenario is approximately equal to the actual price of the income-stabilizing case (which it is supposed to be) except for the slight disturbance of the higher price expected in the two breakdown years. The actual price of Table 7-10 is, however, more than 20 percent higher than the actual price of Table 7-9. The quantity of this uncertainty scenario is somewhat larger than that of Table 7-9, owing to the decreased production of the income stabilizers associated with the higher actual price.

The tendency of some countries to behave as income stabilizers and the energy market participants' unwillingness to commit themselves to capital equipment reflecting the current price may help stabilize the OPEC cartel and also help sustain a price level even higher than the current level.

CONCLUSIONS

The structure of the international petroleum market is oligopolistic. The lack of a unique solution concept for oligopolistic markets makes a bewildering number of price levels plausible outcomes of the oligopolists' market strategies. No uniquely rational behavior can be specified for an individual oligopolist, since the most profitable

Table 7-10. Income Stabilizing Fringe and Uncertainty ($1972)

Year	Market Price	Expected Price	Long-Term Competitive Price
1979	3.67	6.99	3.67
1980	4.20	7.50	4.20
1981	14.00	11.58	4.54
1985	16.36	13.06	5.54
1990	20.07	15.50	7.73

behavior for one seller depends on the response of the others. When the product is relatively homogenous, as is crude oil, and when the demand for the industry's product is inelastic whereas the demand for any single supplier's output is highly elastic, as is also the case for oil, then the range of possible price outcomes is even broader. In such a case the incentive to organize a cartel as well as the incentive to chisel is large.

Conventional oligopoly models are not, however, very useful when dealing with the problem of which suppliers are likely to form a collusive agreement under which circumstances. A more detailed study, story-telling, is required to hypothesize which coalitions are likley to emerge under the different circumstances the industry may have to face. To be able to focus on the likely composition of and the stability of collusive exporter groups, it is necessary to consider the collusive agreements from the point of view of the individual exporters. The heterogeneity of the OPEC membership implies that each member may have a different opinion of what is the "best" policy to pursue. This aspect was indicated in Table 7-2.

A specific analysis of the membership of OPEC is needed to construct a plausible set of intracartel reaction functions. In Chapter Six it was argued that OPEC consists of three different member categories. The members of the first category have the largest reserves of oil, the highest level of excess capacity, the largest financial surpluses, and have demonstrated an ability to work together within a colluding agreement. Saudi Arabia, Kuwait, UAE, Qatar, and Libya were assigned to this category, labeled the "hard core." Iran, Venezuela, Algeria, and Ecuador are presently producing close to their potential and have a strong need for current income. These countries want to continue their present rate of production, but at a higher price. They were labeled the "price pushers." A third category consisting of Iraq, Indonesia, Nigeria, and Gabon was labeled the "expansionist fringe." The members of this group are small relative to the market, produce at a lower production-to-reserves ratio than the price-pushers, and would like to get a somewhat larger share of the market without reducing the price. The expansionist fringe would like the other members of OPEC to accommodate them with a higher market share at the expense of these other members.

There is a number of potential conflicts that might arise with such a membership configuration. Another complicating aspect of the question of cartel stability is the behavior of noncolluding exporters. By allowing noncolluding exporters to follow noncompetitive market strategies, the contribution of such individual noncompetitive strategies to the maintenance of a noncompetitive price-level over time may be assessed.

OPEC has, however, shown an impressive level of discipline by being able to live with considerable excess capacity. The October 1973 cutback of production demonstrated the cohesiveness of the dominating Arab subset of OPEC. The individual exporter governments control the production of domestic oil. Even if the companies still play an important role, their ability to resist producer government initiatives is severely constrained. OPEC as an organization has, however, no instruments available to regulate prices and production rates. OPEC is no more than a forum for discussions and for coordination of the price and production policies of the individual exporting countries. The lack of experience in coordinating production rates, the lack of access to the final consumer markets, the heterogeneity of the producer countries and the increasing government involvement in the marketing of crude, leading possibly to confrontations the exporters are not prepared to face, may make it difficult or impossible for the producer countries to agree on a common tax or price policy.

OPEC has also many of the characteristics of the earlier international commodity cartels that were successful for limited time periods. The lack of formal mechanisms for allocating production and/or profits makes, however, OPEC as vulnerable to the emergence of internal competition as earlier cartels. If OPEC were to follow the pattern set by the nineteen earlier successful cartels reviewed in Chapter Three, then it would likely have a four to six year duration. We would expect, however, to observe some successful post-OPEC cartel-like organizations dominate the international petroleum market for additional four to six year periods. The economic incentive to reestablish OPEC would be stronger than what was the case for any other known cartel.

Our limited experience with high-priced oil/energy and with the implementations of ambitious industrialization programs in the oil exporting countries, makes a discussion based on explicitly stated assumptions the most ambitious analysis of what will happen in the international petroleum market that can realistically be undertaken.

The coefficient and parameter values of this study were estimated from the projections made by OECD, FEA, and other public and private institutions rather than directly from historic observations. In Chapter Six these aspects are discussed. The cost conditions and the price-responsiveness assumed in this study tend to bias the collusive price upwards. The fact that the end 1979 proven recoverable reserves are used as a proxy for ultimately recoverable reserves also biases the collusive price upwards. The opportunity cost of oil as measured by the net present value equivalent of the back-stop technology price will thereby be biased upwards.

The uncertainty with respect to the kind of behavior that will be observed in the international petroleum market, and the desire to evaluate a number of market strategies within the same model framework led to the simple structure of the bathtub model. To be able to perform a number of experiments inexpensively, linear approximations, allowing an analytic expression for the market clearing price to be deducted, were constructed to the relationships we know are much more complex. The linearity assumed introduces biases in the calculations of the various quota systems by underestimating the profits of a cartel member as an independent competitive supplier. The approximations to the long-term monopoly price in the bathtub model do not capture the short-term aspects of monopoly pricing. Short-term exploitation of market power implies a somewhat higher price in the initial years and a lower price in the later years than the bathtub monopoly price path under the same assumptions about the expected price.

The price implications of the various cartel model scenarios are summarized in Table 7-11. The production/exports implications are not listed, as they are sufficiently close to those of Table 7-3, the summary table of the simple model scenarios to make a separate listing unnecessary. It may be argued from Table 7-11 that the present price (of $10.46 in 1975 in 1975 dollars) is lower than the price that would have coincided with the 1975 price of an appropriately weighted average of the price paths indicated in that table. Even if we can observe short-term fluctuations, this analysis indicates that it is unlikely that the average longer-term price will fall below its present level.

**Table 7-11. Summary of Cartel Model Scenarios
(Actual Market Price in 1975 Dollars)**

	1980	*1981*	*1985*	*1990*
Price-Pusher Dominance	11.70		15.52	19.89
Cartel-Core Dominance	11.08		13.85	17.07
OPEC Minus Expansionist Fringe	9.84		12.51	15.73
Cartel Core Monopoly	8.39		11.06	14.35
Reestablishment	6.33	12.36	13.88	17.13
Reestablishment with Income Stabilizing Fringe	6.15	15.63	16.99	20.18
Uncertainty Added	5.58	18.20	21.26	26.08

The OPEC countries will have to continue to live with a considerable level of excess capacity. It seems unlikely that OPEC production will reach more than 80 percent of the existing capacity of 38 MMB/D in the 1980-85 period. The lack of systems for allocating production capacity within OPEC may make the level of excess capacity even higher.

The composition of the three cartel units has been kept constant. Both the composition of the above-mentioned three cartel units and the number of and the characteristics of the cartel units may be changed to produce a number of additional intra-cartel scenarios. A number of additional stories may therefore be told to explore for necessary and sufficient conditions for cartel stability.

The major consumer countries, the importer regions, were assumed to behave as price-takers. Consumer country policies have affected the success of international commodity cartels. By introducing traditional consumer-country policies like tariffs, quotas, taxes, stockpiling, and subsidies some likely necessary and sufficient conditions for cartel stability may be identified under these more complex circumstances.

Even if there is ample room for improvement of the bathtub model in terms of a more realistic representation of the international petroleum market in a world energy market context, the major determinants of the future of the international petroleum market and the relative significance of these determinants as identified in the existing versions of the bathtub model are not expected to change. The major determinants of the future of the international petroleum market are the composition and the cohesiveness of the colluding exporter group, the degree of non-price-taking behavior on the part of the noncolluding exporters, as well as the perceived level of uncertainty with respect to the future level of price in the Persian Gulf.

Appendixes

✳ *Appendix A*

The Market-Clearing Price

A.1 STATIC PERFECT COMPETITION

To write the expression for the market-clearing price in a more compact form, we may simplify the expressions for residual demand and exporter supplies by incorporating all terms that do not involve the current price in a constant term. The constant term of the residual demand function in period t, DL^t, is consequently:

$$DL^t = d^t_1 + \sum_{k=t-K}^{t-1} (-\lambda^k d^k_2 p^k + \lambda^k d_3 (p^k)^2).$$

Equation (5.9) can then be written:

$$D^t = D^t_L - \lambda^t d^t_2 p^t + \lambda^t d_3 (p^t)^2 \tag{A.1}$$

By defining unsubscribed supply coefficients to represent aggregate exporter supplies, and by introducing an analogous constant term, SL^t,

$$SL^t = s0\,(1 - d^t)^t + s2 \sum_{l=t-L}^{t-l} \delta^l p^l \, ,$$

Equation (5.11) can be written in a compact form representing aggregate competitive exporter supplies

113

HUNT LIBRARY
CARNEGIE-MELLON UNIVERSITY

$$S^t = SL^t + \delta^t s2p^t. \tag{A.2}$$

By setting Equation (A.1) equal to (A.2) and solving the system with respect to p^t, the market-equilibrating price in the case of static perfect competition, p_1^t, is found to be:

$$p_1^t = \frac{\lambda^t d_2^t + \delta^t s2}{2\lambda^t d_3} - \sqrt{(\frac{\lambda^t d_2^t + \delta^t s2}{2\lambda^t d_3})^2 - \frac{DL^t - SL^t}{\lambda^t d_3}} \tag{A.3}$$

A.2 STATIC MONOPOLY

The quadratic residual demand function of Equation (5.9) and the monopoly's marginal cost function of the form of the inverse of Equation (5.11) imply that an equation of third degree would have to be solved to find the price at which marginal revenue is equal to marginal cost. To avoid this problem, Equation (5.9) was linearized in the following way. (The "bar" signifies that it is the long-term version of Equation (5.9) that is being considered. That is, no lags are included.)

$$\overline{D}^t = d_1^t - d_2^t p^t + d_3 (p^t)^2 = d_1^t - (d_2^t - d_3 p^t)p^t$$

$$\cong d_1^t - (d_2^t - d_3 p^{t-1})p^t = d_1^t - d_4^t p^t \tag{A.4}$$

$$\overline{D}_L^t = d_1^t - d_4^t p^t$$

As d_3 is small and as p^{t-1} is close to p^t on the "static monopoly" price path or any other "smooth" path, the approximation is fairly accurate. The empirical estimation of d_3 shows, as reported in Chapter Six, that d_3 is less than .04. The linear residual demand function, D_L^t, is to be extrapolated K periods according to the "static monopoly" pricing rule. The resulting marginal revenue of the long-term extrapolated residual demand function is consequently:

$$\overline{MR}^t = d_1^{t+K} - 2d_4^t p^t. \tag{A.5}$$

The long-term version of the competitive supply function is of the form:

$$S^t = s1^t + s2p^t. \tag{A.6}$$

By taking the inverse of this relationship and keeping in mind that quantity is formulated as a function of price rather than price as a function of quantity, the following expression for long-term marginal production costs can be deducted:

$$\overline{MC}^t = \frac{s_1^{t+K} \times d_4^t}{s_2} - \frac{d_1^{t+K} \times d_4^t}{s_2} + \frac{(d_4^t)^2}{s_2} p^t \tag{A.7}$$

By solving for the price that equalizes Equations (A.5) and (A.7), the "static monopoly" price is found to be:

$$p_2^t = \frac{d_1^{t+K} - \dfrac{s_1^{t+K} \times d_4^t}{s_2} + \dfrac{d_1^{t+K} \times d_4^t}{s_2}}{2d_4^t + \dfrac{(d_4^t)^2}{s_2}} \tag{A.8}$$

A.3 INCOME STABILIZATION

By combining Equations (5.9) and (A.4), the short-term linearized residual demand function may be written as:

$$D_L^t = DL^t - \lambda^t d_4^t p^t. \tag{A.9}$$

The market price resulting from equalizing Equations (5.12) and (A.9) is consequently

$$p_3^t = \frac{DL^t}{2\lambda^t d_4^t} - \sqrt{\left(\frac{DL^t}{2\lambda^t d_4^t}\right)^2 - \frac{I^t}{\lambda^t d_4^t}} \tag{A.10}$$

A.4 PRODUCTION STABILIZATION

With a fixed supply path, D^{*t}, the quadratic demand function of Equation (5.9) can be used without causing analytical problems when determining the market price. The resulting market equilibrium price is

$$p_4^t = \frac{d_2^t}{2d_3} - \sqrt{(\frac{d_2^t}{2d_3})^2 - \frac{DL^t - D^{*t}}{\lambda^t d_3}} \qquad (A.11)$$

A.5 TARGET PRICING

When the price is exogenously determined to be TP^t, then the market equilibrating quantity is:

$$S_5^t = DL^t - \lambda^t d_2^t\, TP^t + \lambda^t d_3\, (TP^t)^2 \qquad (A.12)$$

A.6 EXHAUSTIBLE-RESOURCE COMPETITION

The problems we have to face to arrive at an analytic expression for N, the number of periods until exhaustion, when demand grows with time and is also price responsive, can be illustrated by formulating the problem using the continuous version of the simplified long-term linear demand function of Equation (A.4) as a short-term demand function. That is, instantaneous adjustment of the market is assumed. Demand, D^t, is then a linear function of price in the following way:

$$D^t = \alpha_1 e^{\beta t} - \alpha_2 p^t \qquad (A.13)$$

where α_1 and α_2 are appropriately adjusted versions of d_1 and d_4 in Equation (A.4), and β is an assumed rate of market growth. p^t is defined to be

$$p^t = PB\,e^{-r(T-t)}, \text{ and}$$

$$N = T - t.$$

Equation (5.14) then becomes:

$$R = {}_t\!\int^T \alpha_1 e^{\beta t} - \alpha_2 PBe^{-r(T-t)}\, dt. \qquad (A.14)$$

Equation (A.14) can be solved numerically, but it does not seem to be possible to deduct an explicit analytic expression for N or $T-t$ even in this oversimplified case.

As stated earlier, the simulation framework being constructed here is based on explicit analytic expressions for all endogenous variables. Approximating "behavioral" decision rules have been constructed to keep the model framework simple. More complex subroutines to optimize or to solve numerically some segment of the model have been replaced by these behavioral decision rules. The same procedure has been followed to get around the Equation (A.14) problem. By assuming that all the "exhaustible-resource competitors" expect the same rate of growth of the economy, agree on the income elasticity of demand as well as on the price-elasticity of demand, and also expect the market price to grow at the rate of interest as the price is expected to follow the path of the net present value equivalent of the backstop technology price, then all competitors will expect the market to grow at the same rate if it is also assumed that the market price initially was equal to the net present value equivalent of the backstop technology. The expected rate of growth of the market, EG, is simply a function of the elasticity of income, I, the expected rate of growth of the economy, G_E, the price elasticity, E, and the rate of interest, r, as price will increase at the rate of interest on the expected price path,

$$EG = I \times G_E + E \times r. \qquad (A.15)$$

The assumption that the expectations of all competitors with regard to the four variables on the right-hand side of Equation (A.15) are identical will produce an expected growth rate, as can be illustrated in a simple example. If the competitive suppliers expect the economy to grow at a rate of 5 percent per year, assume an income elasticity of one, a rate of interest of 10 percent, and a price-elasticity of −0.1, then they would expect that quantity consumed in each period to increase by 4 percent.

The expected growth rate of the quantity consumed, EG, can be used to solve for the number of periods until exhaustion. N was determined from the following set of equations:

$$R = \sum_{i=t}^{N} D^0 (1 + EG)^t \qquad (A.16)$$

$$R = D^0 \left[\frac{(1 + EG)^N - 1}{EG} \right] \qquad (A.17)$$

$$N = \frac{LOG\left[\frac{EG \times R}{D^0} + 1 \right]}{LOG(1 + EG)} \qquad (A.18)$$

The resulting market price is:

$$p_6^t = \frac{PB}{(1 + r)^{N^t}}$$

(A.19)

How good N^t is as an approximation to the number of periods until exhaustion depends on the level of reserves compared to present production, the price path prior to the p_6^t-path, and how fast the market will adjust to a new price path.

A.7 EXHAUSTIBLE-RESOURCE MONOPOLY

The number of periods until exhaustion is again estimated assuming an expected rate of market growth determined by the net effect of the growth of the economy and the increase in price resulting from the increased size of the market and by the fact that the shadow price of the resource substituted for current marginal production costs is assumed to grow at the rate of interest. It is further assumed that the elasticity of demand in the monopoly case is equal to minus one. In the simple static monopoly case with zero marginal costs then the monopoly price is determined such that the elasticity of demand at the monopoly price is minus one. This does not hold in the case of positive marginal costs or on the Hotelling monopoly path. But it is a fair approximation to the elasticity of demand also in the latter two cases as long as marginal costs are small or the number of periods until exhaustion is large. If marginal costs are small or the number of periods until exhaustion is large, then the monopolist's cost conditions are close to the zero cost case from which the minus one elasticity assumption is deduced.

From these assumptions the monopolist's expected growth of consumption, MG, can be estimated in the case of "exhaustible-resource monopoly pricing" by determining the net change in quantity consumed in each period, resulting from the income effect and the price effect on the monopoly price path. Equation (A.18) then becomes

$$N = \frac{LOG\left[\frac{MG \times R}{D^0} + 1\right]}{LOG\,(1 + MG)}$$

(A.20)

The monopoly price resulting from the equalization of marginal revenue and the net present value equivalent of the ultimate price is:

$$p_7^t = \frac{d_1^t + [PB/(1+r)^{N^t}] \times d_4^t}{2 \times d_4^t}$$

(A.21)

A.8 UNCERTAINTY IN OLIGOPOLISTIC MARKETS

If we assume that p_7^t of Equation (A.21) is the "optimal" expected price, and that the other price options are numbered one to six, then the price that should be charged in period t, p^t, to make the expected price in period t, $E^t(P)$, equal to p_7^t can be found by rearranging the following identity:

$$E^t(P) \equiv \sum_{n=1}^{6} \alpha_n p_n^t + \alpha_7 p_7^t + \beta p^t$$

$$p_7^t \equiv \sum_{n=1}^{6} \alpha_n p_n^t + \alpha_7 p_7^t + \beta p^t$$

(A.22)

$$p^t \equiv p_9^t \equiv \frac{p_7^t(1 - \alpha_7) - \sum_{n=1}^{6} \alpha_n p_n^t}{\beta}$$

Coefficient and Parameter Calues

B.1 THE SIMPLE VERSION

Each of the four importer regions of the bathtub model is described as discussed in Chapter Five and Appendix A, by a set of linear relationships. These relationships are listed in Table B-1. The coefficients $e1_i$, $e2_i$, $m0_i$, $m2_i$, GM_i, $s0_i$, $s2_i$, and d_i^t of these relationships were estimated as described in Chapters Five and Six by fitting a line to a scaled-down version of OECD's and FEA's ("Rest of the World") $3 ($6) and $9 1985 estimates, as

Table B-1. Representation of Importer Regions

Total demand for energy in Region i in Year t, E_i^t (Equation 5.6):

$$E_i^t = e1_1^{t-1} + E_i^{t-1} \times G_i^t - e2_i \sum_{k=t-K}^{t} \lambda^k p^k$$

Market share of oil in Region i in Year t, M_i^t (Equation 5.8):

$$M_i^t = m0_i (1 + GM_i)^t - m2_i \sum_{k=t-K}^{t} \lambda^k p^k$$

Indigenous supplies of oil in Region i in Year t, S_i^t (Equation 5.9):

$$S_i^t = s0_i (1 - d_i^t)^t + s2_i \sum_{k=t-K}^{t} \lambda^k p^k$$

well as the expected path from 1972 to 1985 for indigenous supplies. The coefficient values are listed in Table B-2.

The economic growth assumptions of this study are different from those of the OECD study. The growth assumptions are based on a number of projections that were being made in early 1975. As the range of these projections was substantial, a rather arbitrary average of these projections was chosen as the growth assumptions of this study. In Table B-2 the growth rates assumed are listed in decimal rather than percentage form. That is, in the case of the U.S., the 1974 G^t entry of -0.02 indicate that the U.S. GNP declines by 2 percent in 1974.

The initial intercepts, the $e1_{1972}$'s, were simply estimated to fit the estimated long-term slopes, the $e2$'s, the 1973 growth rates, and the 1973 observed market solution. It was assumed that the 1973 market mix represented the $3 point on the long-term oil/energy relationships.

The coefficients describing indigenous supplies "Rest of the World" were estimated from supply projections made for the individual non-OPEC producers and exporters as reported in Table 6-1 and in Table B-4. In the two cases in which some OPEC members have joined the "Rest of the World" as price-takers, the coefficients were estimated assuming that the end-1974 level of capacity represented the $3, and one-tenth of the end-1974 recoverable reserves represented the $9 level of supply on the long-term competitive supply curve. The coefficient values of the linear supply functions of the exporter coalitions guesstimated as indicated above are reported in Table B-5. The marginal production cost assumptions implicit in these coefficient values can be found by taking the inverse of the assumed competitive supply functions.

The coefficients as reported in Table B-2 summarize the data assumptions underlying the simulations of the "simple" model version. The parameter values common to all importer regions and the simulations of both model versions are listed in Table B-3.

B.2 THE CARTEL VERSION

When estimating the competitive supply functions of the OPEC subunits for the purpose of running the full cartel version of the bathtub model, the end-1974 level of capacity was still assumed to be the $3 competitive supply level, but the $9 competitive supply level was assumed to be only half the level assumed in Table B-5 as a current recoverable-reserves-to-production ratio of twenty rather than of ten was assumed. Otherwise the data assumptions of the cartel version are identical to those of the "simple" version.

Table B-2. Coefficients of Importer Regions

Consumer Region	Key Coefficient Values				Annual Growth and Decline Rates								
					1973	1974	1975	1976	1977	1978	1979	1980-85	1985-90
United States	$e1_{1972}=38.143$ $m0=0.526$ $s0=7.788$	$e2=0.8$ $m2=0.023$ $s2=1.233$	$gm=0.0127$	G^t: d^t:	0.043 .037	-0.02 →	0.01	0.04	0.04	0.05	0.05	0.04	0.04 0
Western Europe	$e1_{1972}=25.37$ $m0=0.678$ $s0^1=0.24$	$e2=0.44$ $m2=0.025$ $s2=0.028$	$gm=0.0057$	G^t: d^t:	0.051 -0.0487	0.02 →	0.02	0.04	0.05	0.06	0.05	0.05	0.05 0
Japan	$e1_{1972}=7.303$ $m0=0.783$ $s0=-0.048$	$e2=0.22$ $m2=0.01$ $s2=0.02$	$gm=0.0016$	G^t: d^t:	0.081 0.056	-0.03 →	0.01	0.04	0.04	0.06	0.07	0.06	0.06 0
"Rest of the World"	$e1_{1972}=18.363$ $m0=0.604$ $s0^2=3.928$ $s0^3=3.928$ $s0^4=8.065$	$e2=0.43$ $m2=0.001$ $s2^2=0.658$ $s2^3=2.758$ $s2^4=5.058$	$gm=0.0014$	G^t: d^{t2}: d^{t3}: d^{t4}:	0.059 -0.2536 -0.1628 -0.1012	0.02 → → →	0.03	0.04	0.05	0.05	0.05	0.05 0.037	0.05 0 0 0

1. Supply from the North Sea is included in indigenous supplies "Rest of the World"
2. Includes all non-OPEC producers and exporters
3. Includes all non-OPEC plus Indonesia, Iraq, Nigeria, and Gabon
4. Includes 3 plus Iran, Algiers, Venezuela, and Ecuador

Table B-3. Parameter Values Common to All Importer Regions

Length of Lag Structure:	10 years		
Distribution of Lags:	$\lambda^t = 0.05$	$\lambda^{t-1} = 0.1$	$\lambda^{t-2} = 0.1$
	$\lambda^{t-3} = 0.1$	$\lambda^{t-4} = 0.1$	$\lambda^{t-5} = 0.1$
	$\lambda^{t-6} = 0.1$	$\lambda^{t-7} = 0.1$	$\lambda^{t-8} = 0.1$
	$\lambda^{t-9} = 0.15$		
Price of Backstop Technology:	1. $15.00 − The simple model version ($-1972)		
	2. $16.00 − The cartel version ($-1972)		

Table B-4. Indigenous Production "Rest of the World" MMB/D

	1973	1980	
		$3	$9
Trinidad and Tobago	0.164	0.20	0.24
Argentina	0.418	0.60	0.24
Bolivia	0.047	0.07	0.10
Colombia	0.199	0.30	0.40
Chile	0.032	0.05	0.05
Peru	0.069	0.10	0.14
Egypt	0.167	0.17	0.20
Syria	0.099	0.20	0.25
Sinai	0.106	0.10	0.10
Turkey	0.066	0.07	0.07
Angola	0.154	0.16	0.20
Tunisia	0.083	0.09	0.10
Congo	0.040	0.04	0.04
Australia	0.419	0.40	0.45
Others and LNG	0.715	0.715	0.715
	2.778	3.265	3.855

Sources: *Petroleum Economist, Petroleum Encyclopedia 1974, Oil and Gas Journal,* various public and private institutions.

In the cartel version the notion of income requirements is introduced. For the purpose of this study, the income requirements of a country in a given year were interpreted to mean the anticipated total imports of the country in that year. The income requirements of the OPEC-membership are represented by a time-dependent equation only,

Table B-5. Coefficient Values of Optimistic Export Scenarios

Coalition	*Key Parameter Values*
All OPEC Countries	$S0 = -10.122$ $S2 = 15.97$ d^t: 73-90 0 $R^a = 489$ 100
OPEC minus Indonesia, Iraq Nigeria and Gabon	$S0 = -10.122$ $S2 = 13.87$ d^t: 73-90 0 $R^a = 418$ 200
Saudi-Arabia, Kuwait, UAE and Libya	$S0 = -14.535$ $S2 = 11.57$ d^t: 73-90 0 $R^a = 327$ 000

Common Parameter Values

Discount Factor:	10 percent
Expected Market Growth:	3 percent
No Lag Structure	

[a]Reserves in million barrel of oil

$$IR_j^t = IRO_j (1 + IG_j^t)^t,$$

where IRO_j is an initial constant and IG_j^t is the rate of growth of OPEC subunit j's income requirements in year t.

As mentioned above the projected total imports path is used as a proxy for the income requirements of the OPEC countries. At the time of this study, there was, however, a bewildering number of estimates of even OPEC's past imports as well as of OPEC's assets abroad. A reasonable estimate of OPEC's 1974 imports seemed to be $35 billion ($-1974) [5]. This figure was broken down into estimates of the total imports of each OPEC subunit [12], and extrapolated at the IG_j^t - growth rates of Table B-6 as $1972 estimates, thereby scaling the estimates up another 16 percent. At the time they were made these estimates looked very optimistic. As it has turned out, they were in fact pessimistic.[1]

The data assumptions for the OPEC membership for the cartel version are summarized in Table B-6.

1. *The Oil and Gas Journal*, 16 February 1976, pp. 42-43.

Table B-6. Coefficient Values of Pessimistic Export Scenarios

Unit	Key Parameter Values
U_1: Iraq, Indonesia, Nigeria, Gabon	$S0 = 4.679$ $S2 = 0.587$ d^t: 73-90 0 R^1: 72700 $IRO^2 = 5763$ $IG_{1973\text{-}1980} = 20$ $IG_{1980\text{-}90} = 10$ $NB^3 = 1$ $AO^4 = -4400$
U_2: Iran, Venezuela, Algeria, Ecuador	$S0 = 10.57$ $S2 = 0.215$ $d^t = 73\text{-}90$ 0 R^1: 91200 $IRO^2 = 7542$ $IG_{1973\text{-}1980} = 20$ $IG_{1980\text{-}90} = 10$ $NB^3 = 1$ $AO^4 = -1690$
U_3: Saudi Arabia, Kuwait, URE & Qatar, Libya	$S0 = 7.675$ $S2 = 4.167$ d^t: 73-90 0 R^1: 327000 $IRO^2 = 8269$ $IG_{1973\text{-}90} = 10$ $NB^3 = 0.5$ $AO^4 = 11800$

Common Parameter Values

Discount Factor	10%
Expected Market Growth	3%
Lag Structure: $\delta^t = 0$ $\delta^{t-1} = 0.25$	$\delta^{t-2} = 0.25$
$\delta^{t-3} = 0.25$	$\delta^{t-4} = 0.25$
Return on Financial Assets Abroad:	5%

1. Reserves in million barrels of oil
2. Income requirements in million dollars
3. Net back factor in dollars per barrel
4. 1973 Level of assets abroad

References

References

CHAPTER ONE: INTRODUCTION

1. M.A. Adelman, *The World Petroleum Market* (Baltimore: Johns Hopkins Press 1972).

2. C. Blitzer, A. Meeraus, A. Stoutjesdijk, "A Dynamic Model of OPEC Trade and Production" (Development Research Center, IBRD, January 1975).

3. R.J. Deam, "World Energy Modeling: Concepts and Methods" (Queen Mary College Energy Research Unit, October 1973).

4. Federal Energy Administration, *Project Independence Report*, November 1974.

5. N.H. Jacoby, *Multinational Oil* (New York: Macmillan Co., 1974).

6. H.A. Houthakker and M. Kennedy, "The World Petroleum Model," *Energy Resources and Economic Growth* (Draft by Data Resources, Inc., for the Ford Energy Policy Project), September 30, 1974.

7. B. Kalymon, "Economic Incentives in OPEC Oil-Pricing Policy," Harvard Business School, April 1975.

8. M. Kennedy, "An Economic Model of the World Oil Market" (doctoral dissertation) Harvard University, August 1974.

9. OECD, *Energy Prospects to 1985* (Paris, 1974).

10. L.A. Rapoport et al., "Modeling of International Long-Range Energy Development and Supplies," proposal to NSF-RANN from Virginia Polytechnic Institute and State University, 1974.

11. J.A. Yager and E.B. Steinberg, *Energy and U.S. Foreign Policy* (Cambridge: Ballinger Publishing Co., 1974).

CHAPTER TWO:
COMPANIES VERSUS EXPORTER GOVERNMENTS

1. M.A. Adelman, *The World Petroleum Market* (Baltimore: Johns Hopkins University Press, 1972).

2. H.R. Alker Jr., L.P. Bloomfield, N. Choucri, *Analyzing Global Interdependence* (Center for International Studies, Massachusetts Institute of Technology, 1974), vol. 2.

3. D.R. Bohi and M. Russel, *U.S. Energy Policy-Alternatives for Security* (Baltimore: Johns Hopkins University Press, 1975).

4. Federal Trade Commission, *The International Petroleum Cartel*, Staff Report to the Federal Trade Commission Submitted to the Subcommittee on Monopoly of the Select Committee on Small Business, U.S. Senate, 82nd Congress, 2d. Session (Committee Print No. 6, 1952).

5. N.J. Jacoby, *Multinational Oil* New York: Macmillan Co., 1974).

6. P.W. MacAvoy, "The Separate Control of Quantity and Price in the Energy Industries" (forthcoming).

7. *Petroleum Press Service*, March 1971, pp. 82-83.

8. *The Petroleum Economist*, December 1974, pp. 459-61.

9. *The Petroleum Economist*, April 1975, pp. 124-26.

10. J.C. Sawhill, Testimony Before the Permanent Subcommittee on Investigations Committee on Government Operations, United States Senate, September 1974.

11. R.B. Stobaugh, Statement Before the Subcommittee on Multinational Corporations of the Committee on Foreign Relations, United States Senate, on "Multinational Petroleum Companies and Foreign Policy," July 25, 1974.

CHAPTER THREE:
THE EXPERIENCE OF PREVIOUS INTERNATIONAL
COMMODITY CARTELS

1. M.A. Adelman, "Politics, Economics, and World Oil" in *Papers and Proceedings of the American Economic Association*, May 1974, pp. 58-67.

2. E. Amid-Hozour, "The Crude Oil Supply: The Middle East, Iran, and the Shell Oil Company," *Tahq Eq.*, Winter-Spring 1972.

3. D.P. Baron, "Limit Pricing, Potential Entry, and Barriers to Entry" *American Economic Review* 63:4 (September 1973).

4. P.T. Bauer and B.S. Yamey, *Markets, Market Control and Marketing Reform* (London: Weidenfeld and Nicolson, 1968).

5. J.R. Behrman, "Monopolistic Cocoa Pricing," *American Journal of Agricultural Economics* 50 (August 1968), pp. 702-719.

6. C.F. Bergsten, "The Threat From the Third World" in *A Reordered World*, edited by R. Cooper (Potomac, November 1973).

7. _____, "The Threat is Real," *Foreign Policy* 14 (Spring 1974), pp. 84-90.

8. _____, "The New Era in World Commodity Markets," *Challenge Magazine* (September/October 1974).

9. _____ , "The Threat From the Third World," *Foreign Policy* 11 (Summer 1973), pp. 102-124.

10. J.C. Burrows, Testimony Before the Subcommittee on Economic Growth of the Joint Economic Committee of the Congress of the United States, July 22, 1974.

11. "Business Gets a New Measure of Bigness," *Business Week*, 27 January 1973.

12. Canada Department of Justice, *Canada and International Cartels* (Ottawa: Canada Department of Justice, 1945).

13. Charles River Associates, "Economic Analysis of the Copper Industry," March 1970.

14. "Japan Acts Against Fiber Cartel" *Chemical Week* 112:29 (January 3, 1973).

15. F. Cheney, *Cartels, Combines and Trusts: A Selected List of References* (Washington, D.C.: Library of Congress, 1944).

16. R.N. Cooper, *A Reordered World* ed. (Potomac, November 1973).

17. J. Devanney, V. Livanos and R. Stewart, "Conference Rate Making and the West Coast of South America," Report No. MITCTL 72-1, Commodity Transportation and Economic Development Laboratory (MIT, January 1972).

18. C. Edwards, *Cartelization in Western Europe* (Washington, D.C.: Bureau of Intelligence and Research, U.S. Department of State, 1964).

19. _____ , *Control of Cartels and Monopolies: An International Comparison* (Dobbs Ferry, N.Y.: Oceana Publications, Inc., 1967).

20. L. Eugwall, *Models of Industrial Structure* (Lexington, Mass.: D.C. Heath, 1973).

21. P.L. Farris, "Market Growth and Concentration Change in U.S. Manufacturing Industries," *Antitrust Bulletin* 18:2 (Summer 1973).

22. Bjarke Fog, "How are Cartel Prices Determined?" *Journal of Industrial Economics*, November 1950.

23. E.T. Grether, "Price Control and Rationing Under the Office of Price Administration," *Journal of Marketing*, vol. 7 (April 1943), pp. 300-318.

24. G.D. Guyer, "Three International Commodity Agreements: The Experience of East Africa," *Econ. Develop. Cult. Change* 21:3 (April 1973), pp. 456-76.

25. W. Hamilton, "The Control of Strategic Materials," *American Economic Review* 34 (June 1944), pp. 261-79.

26. F. Hausmann, "World Oil Control," *Social Research*, 1942.

27. Robert G. Hawkins and Ingo Walter, eds., *The United States and International Markets: Commercial Policy Options in An Age of Control* (Lexington, Mass.: D.C. Heath, 1972).

28. Ervin Hexner, *The International Steel Cartel* (Chapel Hill Press, 1943).

29. _____ , *International Cartels* (London: Pittman, 1946).

30. *International Tea Committee*, Yearly Reports, London.

31. B.L. Johnston, "Potash," *Bureau of Mines Economic Papers* 16 (1933).

32. J.A. Keuler, "The Oil Industry in the World Crisis," *International Cartels* 1 (1939).

33. K.E. Knorr, *World Rubber and its Regulation* (Stanford, Calif.: Stanford University Press, 1945).

34. K.E. Knorr, *Tin Under Control* (Stanford, Calif.: Stanford University Press, 1945).

35. S.D. Krasner, "The Great Oil Sheikdown," *Foreign Policy* 13 (Winter 1973-1974), pp. 123-38.

36. _____, "Oil is the Exception," *Foreign Policy* 14 (Spring 1974), pp. 68-84.

37. W.J. Levy, "An Atlantic-Japanese Energy Policy" in *A Reordered World*, edited by R. Cooper (Potomac, November 1973).

38. H.H. Liebhafsky, "The International Materials Conference in Retrospect," *Quarterly Journal of Economics* 71 (May 1957), pp. 267-88.

39. D. Lynch, "The Concentration of Economic Power," *Indexes of Economic Concentration*, 1946, p. 111.

40. Sir Andrew McFadyean, *History of Rubber Regulation 1934-1943* (London, 1944).

41. J.W. McKil, "The Political Economy of World Petroleum," in *Papers and Proceedings of the American Economic Association* (May 1974), pp. 51-57.

42. Z. Mikdashi, *The Community of Oil Exporting Countries: A Study in Governmental Cooperation* (London: George Allen and Urwin Ltd., 1972).

43. _____, "Collusion Could Work," *Foreign Policy*, 14 (Spring 1974), pp. 57-68.

44. Raymond Mikesell, *Foreign Investment in the Petroleum and Mineral Industries: Case Studies of Investor-Host Country Relations* (Baltimore: Johns Hopkins Press, 1971).

45. _____, "More Third World Cartels Ahead?" *Challenge Magazine*, November/December 1974.

46. J.P. Miller, *Competition, Cartels and Their Regulation* (North-Holland, The Netherlands, 1962).

47. T.H. Moran, "New Deal or Raw Deal in Raw Materials," in *A Reordered World*, edited by R. Cooper (Potomac, November 1973).

48. M. Morrisey and R. Burt, "A Theory of Mineral Discovery: A Note" *Econ. Hist. Rev.*, August 1973.

49. P.C. Newman, *Cartel and Combine: Essays in Monopoly Problems*, (Ridgewood, N.J.: Foreign Studies Institute, 1964).

50. P. O'Brien, "On Commodity Concentration of Exports in Developing Countries," *Economic International* 25:4 (November 1972), pp. 697-717.

51. OECD, "Annual Reports on Developments in the Field of Restrictive Business Practices," 1970.

52. OECD, "Committee of Experts on Restrictive Business Practices," Report, 1972.

53. S. Okita, "National Resource Dependency and Japanese Foreign Policy," *Foreign Affairs* 52 (July 1974), pp. 714-24.

54. S.I. Ornstein, "Determinants of Market Structure," *Southern Economic Journal*, April 1973.

55. Orr and MacAvoy, "Pricing Policies to Promote Cartel Stability," *Economica* 32 (May 1965).

56. G.F. Papanek, "Future Oil Prices and Their Implications for Indonesia," Memorandum, Boston University, 15 March 1973.

57. F.N. Parks and A.H. Hermann, "High Stakes for Cartels, Anti-Trusts, and Governments," *European Business*, Spring 1973.

58. C.E. Pepper, *Air Transportation* (Philadelphia, 1941).

59. G.A. Pollak, "The Economic Consequences of the Energy Crisis," *Foreign Affairs* 52:3 (April 1974).

60. F. Rouhani, *A History of OPEC* (New York: Praeger, 1971).

61. A.E. Sanderson, "Control of Ocean Freight Rates in Foreign Trade," *Trade Promotion Series Report 185*, 1938.

62. Roger Sherman, "Oligopoly: An Empirical Approach," (Lexington, Mass.: D.C. Heath, 1972).

63. Robert Solow, "The Economics of Resources or the Resources of Economics," *The American Economic Review: Papers and Proceedings*, May 1974.

64. G.W. Stocking, *The Potash Industry* (New York, 1931).

65. G.W. Stocking and M.W. Watkins, *Cartels in Action* (Twentieth Century Fund, 1946).

66. _____, *Cartels or Competition* (Twentieth Century Fund, 1948).

67. _____, *Monopoly and Free Enterprise* (Twentieth Century Fund, 1951).

68. G.R. Taylor and I.D. Neir, *The American Railroad Network 1861-1890* (Cambridge: Harvard University Press, 1956).

69. L.C. Tombs, *International Organization of European Air Transport* (London, 1936).

70. John E. Trilton, "The Choice of Trading Partners: An Analysis of International Trade in Aluminum, Bauxite, Copper, Lead, Manganese, Tin and Zinc," *Yale Economic Essays* 6 (Fall 1966).

71. United Nations, "Restrictive Business Practices," 1973 Sales No. E. 73. II.D.8.

72. U.S. Federal Maritime Commission, "Investigation of Ocean Rate Structures in the Trade Between U.S. North Atlantic Ports and Ports in the U.K. and Eire-North Atlantic U.K. Freight Conference, Agreement 7100, and North Atlantic W/B Freight Association, Agreement 5850," Docket No. 65-45, 14 July 1967.

73. U.S. Federal Trade Commission, "The International Petroleum Cartel," Committee Print No. 6, 1952.

74. _____, "International Steel Cartels," 1948.

75. _____, "International Cartels in the Alkali Industry," 1950.

76. B. Varon and K. Takeocki, "Developing Countries and Non-Fuel Minerals," *Foreign Affairs* 52 (April 1974), pp. 497-509.

77. Raymond Vernon, "Foreign Enterprises and Developing Nations in the Raw Materials Industries," *The American Economic Review* 61:2 (May 1970).

78. Donald H. Wallace, *Market Control in the Aluminum Industry* (Cambridge, Mass.: 1937).

79. C.R. Whittlerey, *Government Control of Crude Rubber* (Princeton, 1931).

80. Kurt Willie, "The International Sugar Regime," *American Political Science Review*, October 1939.

81. H. Wood, "No Agreement on Cocoa," *Cartel* 14:1 (January 1964), pp. 17-23.

82. _____ , "The World Coffee Economy," *Cartel* 13:4 (October 1963), pp. 160-68.

83. "The World Sugar Economy," *Cartel* 13:3 (July 1963), pp. 124-29.

84. Yuan-Li Wu, *Raw Material Supply in a Multipolar World* (New York: Crane, Russak and Co., for the National Strategy Center, Inc., 1973).

Marginal References

85. H. Arndt, "Some Fundamental Questions of Concentration Policy," *German Economic Review* 10:2 (1972), pp. 101-115.

86. D. Burn and B. Epstein, *Realities of Free Trade: Two Industry Studies* (University of Toronto Press, 1972).

87. C. Canenbley, "Price Discrimination and EEC Cartel Law: A Review of the Kodak Decision of the Commission of the European Communities," *Antitrust Bulletin* 17:1 (Spring 1972), pp. 269-81.

88. A.J. Cordell, "The Brazilian Soluble Coffee Problem: Reply," *Quarterly Review of Economics and Business* 82:326 (Autumn 1972), pp. 629-40.

89. C.W. Corssmit, "The Oil Import Quota System: Time for Reform," *Intermountain Economic Review* 2:2 (Fall 1971), pp. 35-46.

90. D.M. Etherinton, "An International Tea Trade Policy for East Africa: An Exercise in Oligopolistic Reasoning," *Food Research Institute Study* 11:1 (1972), pp. 89-108.

91. F. Fama and A. Laffer, "The Number of Firms and Competition," *American Economic Review* 62:4 (September 1972).

92. C.A. Gasoliba, "Estudio Economico de las Industrias Carnicas" (Spain: Banca Catalana, 1972).

93. B. Griffiths, "The Welfare Cost of U.K. Clearing Banks Cartel," *Journal of Money, Credit and Banking* 4:2 (May 1972), pp. 227-44.

94. F.O. Grogan, ed., *International Trade in Temperate Zone Products* (Edinburgh: Oliver and Boyd University of New Castle upon Tyne, 1972).

95. John A. Guthrie, *An Economic Analysis of the Pulp and Paper Industry* (Pullman, Washington: Washington State University Press, 1972), p. 235.

96. P. Hanson, "Soviet Imports of Primary Products: A Case Study of Cocoa," *Soviet Stud.* 23:1 (July 1971), pp. 59-77.

97. R. Hellman, *Government Competition in the Electric Utility Industry: A Theoretical and Empirical Study* (Praeger Special Studies in U.S. Economic and Social Development, 1972).

98. J.A. Hopkins, "Conglomerate Growth in Agriculture—Some Policy Implications," *Agricultural Finance Review* 33 (July 1972), pp. 15-21.

99. P.B. Kenen, "1972 Report of President's Council of Economic Advisors: International Aspects."

100. K. Kojima, "Nontariff Barrier; to Japanese Trade," *Hitosubashi Journal of Economics* 13:1 (June 1972), pp. 1-39.

101. B.J. Linder and A. Sarkar, "Pakistan's Monopolies and Restrictive Practices Ordinance," *Antitrust Bulletin* 16:3 (Fall 1971), pp. 569-83.

102. D.L. Losman, "The Effects of Economic Boycotts," *Lloyd's Bank Review* 106 (October 1972), pp. 27-41.

103. C. Marfel, "A Guide to the Literature on the Measurement of Indus-

trial Concentration in the Post-War Period," *Z. National Ø Kon* 31:3-4 (1971), pp. 483-506.

104. J. Nash, "The Devil in Bolivia's Nationalized Tin Mines," *Sci. Soc.* 36:2 (Summer 1972), pp. 221-33.

105. L.H. Officer, "Discrimination in the International Transportation Industry," *Western Economics* 10 (June 1972), pp. 170-81.

106. J.K. Olayemi, "Technology in the Western Nigeria Cocoa Industry: An Empirical Study," *Economic Bulletin of Ghana* 2:2 (1972), pp. 47-60.

107. C.K. Rowley, *Steel and Public Policy* (London, New York: McGraw-Hill, 1971).

108. H. Roy, "Some Observations on New International Tea Agreement," *Economic Affairs* 17:8 (August 1972), pp. 386-94.

109. R.J. Sampson, "Inherent Advantages Under Regulation," *American Economic Review* 62:2 (May 1972).

110. G.K. Sarkar, *The World Tea Economy* (New York: Oxford University Press, 1972).

111. M.C. Sawyer, "Concentration in the British Manufacturing Industry: A Reply," *Oxford Economic Papers* 24:3 (November 1972), pp. 438-47.

112. A. Schmitz, "Tariffs and Declining Cost Industries," *Economica*, September 1972, pp. 419-26.

113. L.W. Stern, "Antitrust Implications of a Sociological Interpretation of Competition, Conflict and Cooperation in the Market-Place," *Antitrust Bulletin* 16:3 (Fall 1971), pp. 509-530.

114. J.E. Tilton, *International Diffusion of Technology: The Case of Semi-Conductors* (Washington, D.C.: The Brookings Institution, 1971).

115. W.G. Tyler, "The Brazilian Soluble Coffee Problem: Comment," *Quarterly Review of Economics and Business* 12:3 (Autumn 1972), pp. 99-102.

116. R.I. Williams, "Jamaican Coffee Supply, 1953-68: An Exploratory Study," *Soc. Econ. Stud.* 21:1 (March 1972), pp. 90-103.

117. I. Yamazawa, "Industry Growth and Foreign Trade—A Study of Japan's Steel Industry," *Hitotsubaski Journal of Economics* 12:2 (February 1972), pp. 41-59.

CHAPTER FOUR:
CARTEL THEORY AND ITS RELEVANCE TO
THE INTERNATIONAL PETROLEUM MARKET

1. J.S. Bain, *Barriers to New Competition: Their Character and Consequences in Manufacturing Industries* (Cambridge, Mass., 1956).

2. D. Baron, "Limit Pricing, Potential Entry, and Barriers to Entry," *American Economic Review* 63 (September 1973), pp. 660-74.

3. J. Bhagwati, "Oligopoly Theory, Entry Prevention and Growth," *Oxford Economic Papers* 22 (November 1970), pp. 297-310.

4. R.L. Bishop, "Duopoly: Collusion or Warfare," *American Economic Review* 59 (September 1960), pp. 933-61.

5. E.H. Chamberlin, *The Theory of Monopolistic Competition* (Cambridge, Mass., 1933).

6. A. Cournot, *Researches into the Mathematical Principles of the Theory of Wealth*, translated by N.T. Bacon (New York, 1897, reprinted 1929).

7. W. Fellner, *Competition Among the Few* (New York: Knopf, 1949).

8. D. Gaskins, "Dynamic Limit Pricing: Optimal Pricing Under Threat of Entry," *Journal of Economic Theory* 3 (September 1971), pp. 306-22.

9. R.B. Helfebower and G.W. Stocking, eds. (for the American Economic Association), *Readings in Industrial Organization and Public Policy* (Homewood, Ill.: Richard D. Irwin, Inc., 1958).

10. P.L. Joskow, "Firm Decision-Making Processes and Oligopoly Theory," *American Economic Association*, May 1975.

11. M. Kamien and N. Schwartz, "Limit Pricing and Uncertain Entry," *Econometrica* 39 (May 1971), pp. 441-54.

12. F. Modigliani, "New Developments on the Oligopoly Front," *Journal of Political Economy* 66 (June 1958), pp. 215-32.

13. R. Nelson and S. Winter, "Neoclassical Versus Evolutionary Theories of Economics Growth; Critique and Prospectus," *Economic Journal*, 1974, pp. 886-905.

14. D. Orr and P.W. MacAvoy, "Price Strategies to Promote Cartel Stability," *Economica* 32 (May 1965), pp. 186-97.

15. R. Schmalensee, *Applied Microeconomics* (Holden-Day, Inc., 1973).

16. A.M. Schreiber, ed., *Corporate Simulation Models* (Seattle, 1970).

17. M. Shubik, "Oligopoly Theory, Communication, and Information," *American Economic Association*, May 1975.

18. G.J. Stigler, "A Theory of Oligopoly," *Journal of Political Economy* 72 (February 1964), pp. 44-61.

19. G.J. Stigler and K.E. Boulding, eds. (for the American Economic Association), *Readings in Price Theory* (Homewood, Ill.: Richard D. Irwin, Inc., 1952).

CHAPTER FIVE:
THE MODEL REPRESENTATION OF THE
INTERNATIONAL PETROLEUM MARKET

1. M.A. Adelman, "Economics of Exploration for Petroleum and Other Minerals," *Geoexploration* 8 (1970).

2. Federal Energy Administration Report, *Project Independence Report*, November 1974.

3. R.L. Gordon, "A Reinterpretation of the Pure Theory of Exhaustion," *Journal of Political Economy*, 1967.

4. _____, "Mythology and Reality in Energy Policy," in *Energy Policy*, September 1974.

5. O.C. Herfindahl, "Depletion and Economic Theory," in *Extractive Resources and Taxation*, edited by M. Gaffney (University of Wisconsin Press, 1967), pp. 63-90.

6. H. Hotelling, "The Economics of Exhaustible Resources," *Journal of Political Economy* 39 (April 1931), pp. 137-175.

7. W.D. Nordhaus, "The Allocation of Energy Resources," in *Brookings Papers on Economic Activity* 3 (1973).

8. OECD, "Energy Prospects to 1985" (Paris, 1974).

CHAPTER SIX:
ANALYTICAL UNITS, DATA, AND THE FRAMEWORK
OF THE STUDY
AND APPENDIX B:
COEFFICIENT AND PARAMETER VALUES

1. M.A. Adelman, *The World Petroleum Market* (Baltimore: Johns Hopkins Press, 1972).

2. H.R. Alker Jr., L.P. Bloomfield, N. Choucri, *Analyzing Global Interdependence* (Center for International Studies, Massachusetts Institute of Technology, 1974), vol. 2.

3. D.R. Bohi and M. Russel, *U.S. Energy Policy—Alternatives for Security* (Baltimore: Johns Hopkins University Press, 1975).

4. Bureau of Mines, "International Petroleum Annual 1973," U.S. Department of the Interior.

5. First National City Bank, *Monthly Economic Letter*, June 1975.

6. Federal Energy Administration Report, *Project Independence Report*, November 1974.

7. M. Kennedy, *An Economic Model of the World Oil Market* (doctoral dissertation), Harvard University, August 1974.

8. Massachusetts Institute of Technology Energy Laboratory, Policy Study Group, "The FEA Project Independence Report: An Analytical Assessment and Evaluation," forthcoming.

9. W.D. Nordhaus, "The Allocation of Energy Resources," in *Brookings Papers on Economic Activity* 3 (1973).

10. OECD, "Energy Prospects to 1985" (Paris, 1974).

11. *Petroleum Economist.*

12. *Petroleum Economist*, March 1975, pp. 84-86.

13. R.B. Stobaugh, Statement Before the Subcommittee on Multinational Corporations of the Committee on Foreign Relations, U.S. Senate, on "Multinational Petroleum Companies and Foreign Policy," July 25, 1974.

Index

Abu Dhabi: market experience, 23; nationalization, 16
Adelman, M.A., 6
Algeria: market experience, 23; monopoly price, 90; politics, 22; price pusher, 107
aluminum: cartel overview, 33
Arab League, 12

Bergsten, C. Fred, 33, 34
Bishop, R.L., 48
Blitzer, C., et al., 6
Brazil: revised potential, 77
Brunei-Malaysia, 76
Burrows, James, C., 30

Canada, 2, 77
cartel: and bathtub simulations model, 94; behavior, 49; characteristics, 27–30; collusive arrangements, 23; commodity overview, 30–36; comparative analysis, 41; core dominance, 100; definition, 47; discipline and bathtub model, 86; efficiency, 43; exporter fringe, 76; initial movements, 22; and monopoly pricing, 7; nature of, 25; price advantage over monopoly, 93; pricing, 67; price-pushers, 80; price strategy, 70; production and profit allocation, 44; theory, 5
Chile: CIPEC, 36
CIPEC: organization, 36
coffee: cartel overview, 40

collusive agreements, 50
commodities: cartel organization, 30–35
competition: equilibrium price, 104; exhaustible-resource, 65
concessions, 10; and nationalism, 11; producer government, 21
Congo: COPEC, 36
consumers: concentration level, 44; regions of in bathtub model, 74
copper: cartel overview, 35, 38

Deam, R.J., 6
demand: elasticity defined, 75; residual world, 4
depletion: and bargaining ability, 78; notion of, 3
diamonds: cartel overview, 40
distribution: and role of cartel, 43

Ecuador, 81; price pusher, 107
efficiency: cartel characteristics, 28; and cartel organization, 31–41
Egypt: politics, 22
embargo, 23; and politics, 80
energy: consumption in target year, 75; demand function, 58
ENI: joint-venture agreements, 14
ERAP, 15
Europe: imports, 1
exporters: bargaining power, 13, 17; and cartel characteristics, 28; cohesiveness of colluding groups, 110; collusive arrangements, 94; decision-

About the Author

Dr. Eckbo is associated with the Energy Policy Group of the MIT Energy Laboratory and the Center of Applied Research of the Norwegian School of Economics and Business Administration. After completing his graduate work at Sloan School of Management, MIT, he has been conducting research and consulting in the area of supply of natural resources and the organization of international commodity markets in a broader international management context.

DATE DUE